Managing Risks in Digital Transformation

Navigate the modern landscape of digital threats with the help of real-world examples and use cases

Ashish Kumar

Shashank Kumar

Abbas Kudrati

BIRMINGHAM—MUMBAI

Managing Risks in Digital Transformation

Group Product Manager: Mohd. Riyan Khan

Publishing Product Manager: Prachi Sawant

Senior Editor: Arun Nadar

Content Development Editor: Sujata Tripathi

Technical Editor: Nithik Cheruvakodan

Copy Editor: Safis Editing

Project Coordinator: Ashwin Dinesh Kharwa

Proofreader: Safis Editing

Indexer: Subalakshmi Govindhan

Production Designer: Shankar Kalbhor

Marketing Coordinator: Marylou Dmello

First published: April 2023

Production reference: 1160323

Published by Packt Publishing Ltd.
Livery Place
35 Livery Street
Birmingham
B3 2PB, UK.

ISBN 978-1-80324-651-2

www.packtpub.com

Dedicated to all security and IT helpdesk professionals who made working from home possible during the pandemic.

– Ashish Kumar

To my kid, Shrina, for her infectious curiosity and optimism.

– Shashank Kumar

To all my mentors who inspired me to achieve more in my professional and personal life. Thank you, Ruzan Khambhata (Wizz o Tech), Govind Kaveri (BHA), M. K. Srinivasan and Navin Gabrial (WebXL Asia), Ahmed Buhazza (eGov Bahrain), Srikant Ranganathan (E&Y), Tom Gamali (NBK), Anthony Stevens (KPMG), and Avinash Lotke, Mandana Javaheri, and Sian John (Microsoft). Thank you to my Toastmasters mentors, Mohd Isa, Khalid Alqoud, Shaukat Lokhandwala, and Zulfiqar Ali. Your guidance and coaching have helped me to become a better person, professional, speaker and leader today.

– Abbas Kudrati

Foreword

I have had a long and impactful career in business continuity, disaster recovery, cybersecurity, and crisis management, which has enabled me to operate holistically to understand the threat landscape and how this understanding can be proactively enabled to deliver effective resilience. I have been at the frontline of many major incidents, including IT failures, cyberattacks, data breaches, and fraud.

This all led me to write my first book, *Effective Crisis Management*, where I looked back over the last 25 years of my career to explore the evolving threat landscape – whetherfacing it's economic, geopolitical, environmental, or technological risks, organizations need innovative ways to understand and navigate the array of threats that they face on a daily basis.

It is this shared passion and enthusiasm for risk and resilience that got me excited about writing the foreword to this book. The authors have really brought to life the challenges that so many organizations face today.

The combined experience and backgrounds of Ashish Kumar, Abbas Kudrati, and Shashank Kumar make for compelling reading, as they share their deep insights and strategic thinking. Their leadership and subject matter expertise shine through each chapter.

The only true certainty is that change is constant. What perhaps sets the current landscape apart from previous years is the speed and scale of change that we are experiencing, which is at an unprecedented level.

I really enjoyed how the authors have explored some of these issues, learning how certain events have led to seismic shifts, not just in our level of tolerance to risk but also in the public perception of risk. In particular, the authors take us on a journey and explore the rich history, showing how the correlation between major incidents and distrust has led to wholesale regulatory change.

A prime example of this is the 2016 **General Data Protection Regulation** (**GDPR**) – one of the heaviest debated regulations in EU history – which holds organizations to account on how they collect, process, and analyze personal data. In the digital era, this has led many countries to adopt similar regulatory requirements and standards for effective privacy and security controls.

This book will appeal to a wide range of risk, resilience, and security practitioners – from those with strategic accountability and ownership for establishing the risk appetite and frameworks to those with operational responsibility for identifying and managing risk. The book is written in a format that is easy to digest and understand, with a myriad of practical examples, case studies, and some eye-watering statistics to bring the subject to life. This application of real-world knowledge and experience makes the authors' insights so valuable.

You can expect practical hints and tips about how to navigate the evolving risk and compliance threat landscape that will provide longevity in the adoption of successful risk frameworks and contingencies. Some of the key outcomes include developing a view on zero trust architecture, understanding the insider threat landscape, and getting contextual frameworks for governance and regulatory risk management. This book is a must-read for anyone with an interest in risk and compliance.

Sarah Armstrong-Smith

Chief Security Advisor, EMEA

Microsoft

Contributors

About the authors

Ashish Kumar is a principal product manager at Microsoft. He has over 25 years of experience in networking, infrastructure, and cloud to cybersecurity at Microsoft, TCS, and HCL, and he has engaged with over 2,000 customers, representing their viewpoints with regard to the current times. He has played various roles in leading cybersecurity discussions with CISOs, IT leaders, and businesses for over 10 years. He has actively learned about and gained insights into digital risk in his regular discussions with over 100 chief risk officers and CISOs in the last two years as part of his engineering role. He holds various certifications, such as MCSE, CCNP, ISO Auditor, Cloud, and Power Platform certifications.

Shashank Kumar is a cybersecurity practitioner, longtime regulatory risk enthusiast, and principal product manager for Microsoft Purview Compliance products. He works closely with some of the world's largest corporations to help them understand their current and future cybersecurity risks and solve them, through new features or products from Microsoft Data Security Product Group. He has exhaustive experience in helping Fortune 500 companies design their security and compliance management strategies, and he is also a frequent contributor on product management and cybersecurity forums.

Abbas Kudrati is a longtime cybersecurity practitioner and CISO and is Microsoft Asia's chief cybersecurity advisor. In addition to his work at Microsoft, he serves as an executive advisor to Deakin University, HITRUST, EC-Council, and several security and technology start-ups. He supports the broader security community through his work with ISACA chapters and student mentorship. He is the technical editor of various books and the bestselling author of books such as *Threat Hunting in the Cloud* and *Zero Trust Journey Across the Digital Estate*. He is also a part-time professor of practice at La Trobe University and has been a keynote speaker on zero trust, cybersecurity, cloud security, governance, risk, and compliance.

About the reviewers

Hardik Kiran Mehta has contributed to information security, risk, and compliance for more than 17 years, specializing in global privacy laws, NIST, ISO, and risk management. He has performed mobile forensic investigations for both law enforcement and the intelligence community in support of the US federal government. He received a BE in computer engineering from Mumbai University and a master's in computer science from Stevens Institute of Technology. He is currently employed by Microsoft as director of security, risk, and compliance at their headquarters in Redmond, USA. He has received various accolades and awards from the government of Australia, SEC, and ISACA.

I'd like to thank my family and friends who understand the time and commitment it takes to contribute actively to information security, risk, and compliance, which is constantly changing. Working in this field would not be possible without the supportive security, risk, and compliance community that has developed over the last several years. Thank you to all of the trailblazers who make this field an exciting place to work each and every day.

Salah Eddine MAHRACH is a leading, trusted GRC professional with more than 19 years of business experience in the banking, financial services, and energy sectors. Leveraging years of field experience, he has succeeded in implementing risk management, audit practices, and structures for numerous corporations, aligning with multiple international standards and frameworks. He has an interest in IT governance, risk management, IT audits, investigations, business continuity management, and regulatory compliance, which he likes to share as a volunteer in working groups or as a speaker at conferences. He is also one of the founders of the ISACA Moroccan chapter. Salah Eddine holds a number of professional certifications, including CISA, CRISC, CDPSE, and COBIT 2019 Foundation.

I am thankful to my family, especially my wife and my kids, for their support and for tolerating my busy schedule and still standing by my side.

I am deeply indebted to ISACA and its community for making available valuable knowledge and access to an outstanding professional network.

Table of Contents

Part 2: Risk Redefined at Work

7

Modern Collaboration and Risk Amplification   85

8

Insider Risk and Impact   97

9

Real Examples and Scenarios   111

13

The Role of Data and Privacy in Risk Management   153

Part 3: The Future

14

Remote Work and the Virtual Workforce   169

15

Automation and Virtual Humans   179

16

The Role of AI in Managing Future Lockdowns   191

Preface

In a world increasingly dominated by technology, digital transformation has become a critical driver of growth and competitiveness for organizations of all sizes and industries. With the rise of cloud computing, mobile devices, the Internet of Things, and other innovative technologies, companies can collect, store, and process vast amounts of data in real time, opening up new opportunities for business transformation, increased efficiency, and new revenue streams.

However, as organizations embrace digital transformation, they also face new and complex risks. The modern landscape of digital threats is constantly evolving, and companies must be vigilant in their efforts to protect sensitive information, systems, and infrastructure. From cyber-attacks and data breaches to security failures and other forms of digital risk, companies must have a deep understanding of the challenges they face, as well as the tools and strategies needed to manage these risks effectively.

This book is designed to help organizations navigate the complex and rapidly changing world of digital risk. Through a series of real-world examples and use cases, we will explore the key challenges and risks associated with digital transformation and provide practical insights and strategies for managing these risks. Our goal is to help organizations understand the modern landscape of digital threats and provide them with the tools and knowledge they need to succeed in a digital world.

Whether you are a business leader, security professional, or IT manager, this book will provide you with a comprehensive overview of the challenges and risks associated with digital transformation, as well as the strategies and best practices you need to manage these risks effectively. With a focus on real-world examples and practical insights, this book is an essential resource for anyone looking to navigate the modern landscape of digital risk.

For us, this book is an attempt to build a conversation among the large fraternity of business, technology, and cybersecurity enthusiasts, leaders, and practitioners. Please feel free to reach out to us:

Abbas Kudrati – @askudrati (Twitter) and `https://www.linkedin.com/in/akudrati/`

Ashish Kumar – `linkedin.com/in/ashishkadhikari`

Shashank Kumar – @Shshank (Twitter) and `linkedin.com/shashank1kumar`

Who this book is for

Managing Risks in Digital Transformation is broadly focused on assisting three categories of readers—first, those who own a business of any size and are planning to scale it; second, those who are leading business and technology charters in large companies or institutions; and third, those who are academically or disciplinarily targeting cybersecurity and risk management as an area of practice. Essentially, this book is for business leaders, board members, small and medium business owners, and professionals working in IT, risk, governance, compliance, and legal domains. It is designed to help technology leaders such as chief digital officers, chief privacy officers, chief risk officers, CISOs, and CIOs, and will help students and cybersecurity enthusiasts to develop a basic awareness of risks to navigate the digital threat landscape.

What this book covers

Chapter 1, Invisible Digitization Tsunami, gives the reader a view into the domains of human work and personal life a few decades back and how fast they have changed.

Chapter 2, Going Digital, provides insight into how our lives, both personal and professional, are saturated with technology. From digital assistants to smartwatches, this chapter discusses how we as humans are becoming increasingly dependent on technology.

Chapter 3, Visible and Invisible Risks, identifies the visible and invisible risks involved in real-life scenarios, from browsing the internet to using an application on our mobile phones.

Chapter 4, Remote Working and the Element of Trust, focuses on the topic of remote working, which is now widespread due to COVID-19. It examines the history and concept of **working from home** (**WFH**), the impact of the pandemic, views from various industries, and the risks it presents to organizations.

Chapter 5, Emergence of Zero Trust and Risk Equation, examines how the emergence of zero trust security architecture and risk equation reflects a paradigm shift in cybersecurity. Zero trust emphasizes the need to verify every request and restrict access to resources and aims to balance the costs of security measures against potential losses from cyber threats.

Chapter 6, The Human Risk at the Workplace, goes through the types of risks in the workplace, who they involve, and their repercussions, profiling employees through the lens of academic research on digital risk and live examples. The chapter also illustrates the ways in which each distinct persona is susceptible to digitally risky behavior.

Chapter 7, Modern Collaboration and Risk Amplification, tracks the recent evolution of collaboration in enterprises and institutional workspaces and the implications it has for both employees and management.

Chapter 8, Insider Risk and Impact, offers a qualitative and quantitative approach to evaluating and understanding the implications of insider risk.

Chapter 9, Real Examples and Scenarios, contains four real stories from the corporate world, anonymized but carrying actual details of the way a large data breach or business impact panned out due to the risky behavior of an employee.

Chapter 10, Cyberwarfare, elaborates on the concept of war and cyberwarfare. War refers to an armed conflict between countries or entities, while warfare refers to the tactics used to win. Cyberwarfare is a new form of warfare that uses technology to attack an enemy and inflict damage on physical objects. The chapter will explore the impact of cyberwarfare on countries and organizations and examine the various actors involved, including nation states and cybercriminals.

Chapter 11, An Introduction to Regulatory Risks, contextualizes digital regulatory risk for an average reader. The reader is introduced to a few frameworks that should help them understand the need for the regulations in question and the implications of regulatory risk.

Chapter 12, The Evolution of Risk and Compliance Management, follows the evolution of modern compliance management as a discipline from the common origins of risk management. It also takes readers through a timeline of corporate scandals and scams and correlates those with the development of regulatory frameworks from governments and institutions in response.

Chapter 13, The Role of Data and Privacy in Risk Management, establishes the size of the issue when it comes to enterprise data and introduces readers to the need for companies to responsibly retain or delete their data in the context of modern privacy regulations.

Chapter 14, Remote Work and the Virtual Workforce, discusses the relationship between remote working and AI. The authors believes that AI is changing the nature of work, leading to a redefinition of work and the emergence of new categories of workers known as "work beings". The authors raises concerns about the impact of this shift on human social connections and well-being.

Chapter 15, Automation and Virtual Humans, explores the idea of human and work beings in the context of technological advancements such as AI and automation. "Work beings" refers to new forms of workers such as robots or avatars that can do tasks previously performed by humans. The chapter also explores how automation and AI will shape the presence of work beings in the workforce and covers topics such as the current state of automation and the development of chatbots and digital humans.

Chapter 16, The Role of AI in Managing Future Lockdowns, looks ahead at the next two decades. The number of internet-connected devices is expected to surpass the number of humans, leading to significant changes in the form and interface of digital devices. The use of AI in these devices will impact human interactions and relationships, creating new habits and posing new risks, such as digital lockdowns that may disrupt electricity and internet connectivity. Laws and regulations are needed to prevent and mitigate these risks and ensure ethical practices in AI technology companies.

Download the color images

We also provide a PDF file that has color images of the screenshots and diagrams used in this book. You can download it here: `https://packt.link/MWKJk`.

Conventions used

There are a number of text conventions used throughout this book.

`Code in text`: Indicates code words in text, database table names, folder names, filenames, file extensions, pathnames, dummy URLs, user input, and Twitter handles. Here is an example: "Create a new file named `argocd-rbac-cm.yaml` in the same location as `argocd-cm.yaml`."

Bold: Indicates a new term, an important word, or words that you see onscreen. For instance, words in menus or dialog boxes appear in **bold**. Here is an example: "You can use the UI by navigating to the **User-Info** section."

> **Tips or important notes**
> Appear like this.

Get in touch

Feedback from our readers is always welcome.

General feedback: If you have questions about any aspect of this book, email us at `customercare@packtpub.com` and mention the book title in the subject of your message.

Errata: Although we have taken every care to ensure the accuracy of our content, mistakes do happen. If you have found a mistake in this book, we would be grateful if you would report this to us. Please visit `www.packtpub.com/support/errata` and fill in the form.

Piracy: If you come across any illegal copies of our works in any form on the internet, we would be grateful if you would provide us with the location address or website name. Please contact us at `copyright@packt.com` with a link to the material.

If you are interested in becoming an author: If there is a topic that you have expertise in and you are interested in either writing or contributing to a book, please visit `authors.packtpub.com`.

Share Your Thoughts

Once you've read *Managing Risks in Digital Transformation*, we'd love to hear your thoughts! Scan the QR code below to go straight to the Amazon review page for this book and share your feedback.

https://packt.link/r/1803246510

Your review is important to us and the tech community and will help us make sure we're delivering excellent quality content.

Download a free PDF copy of this book

Thanks for purchasing this book!

Do you like to read on the go but are unable to carry your print books everywhere? Is your eBook purchase not compatible with the device of your choice?

Don't worry, now with every Packt book you get a DRM-free PDF version of that book at no cost.

Read anywhere, any place, on any device. Search, copy, and paste code from your favorite technical books directly into your application.

The perks don't stop there, you can get exclusive access to discounts, newsletters, and great free content in your inbox daily

Follow these simple steps to get the benefits:

1. Scan the QR code or visit the link below

https://packt.link/free-ebook/9781803246512

2. Submit your proof of purchase

3. That's it! We'll send your free PDF and other benefits to your email directly

Part 1: Invisible Digitization Tsunami

This part covers the impact of digitization on modern society. It begins with a comparison of the "good old days" and our current era and then moves on to discuss the ubiquitous presence of digital technology. This part also explores the risks associated with digitization, both visible and invisible, and the rise of remote work. Finally, it delves into the concept of zero trust and how that affects the overall risk equation. The aim of this part is to provide an overview of the changes brought about by digitization and the associated risks.

This part of the book contains the following chapters:

- *Chapter 1, Invisible Digitization Tsunami*
- *Chapter 2, Going Digital*
- *Chapter 3, Visible and Invisible Risks*
- *Chapter 4, Remote Working and the Element of Trust*
- *Chapter 5, The Emergence of Zero Trust and Risk Equation*

Invisible Digitization Tsunami

It's a bright day in 2023, and most humans on the planet are acclimating to the new normal after the pandemic that changed the way we work and live, while a few months back, Amazon founder Jeff Bezos took the first civil flight to space, creating a milestone. The world around us is changing fast. The human population in 2022 was around 8 billion, and most of us had a mobile phone; the count of phones is hovering around the 10 billion mark. What's moving faster than the human and mobile population is the count of internet-connected smart devices, also known as IoT devices. Today, they are found in cars, smart homes, and industrial devices and they number 13.5+ billion at the time of writing. That totals up to 24 billion internet-connected devices between 8 billion humans.

On a personal front, I think the number of virtual assistant devices, such as Amazon devices, will beat the estimate of 90 million for 2022. In 2021, Amazon sold close to 55 million devices. Sometimes, you may wonder why we did not have such innovation a few years back. I remember the shift humans made from the once-dominant Sony Walkman to CDs, and then to mass storage devices, such as the Apple iPod and the MP3 format. I owned an iPod, and it was a cool product that Apple launched in 2001; it got its last update at some point in 2014. While at its peak Apple sold close to 51 million iPods, it still missed the innovation spotted by Amazon – voice command technology. Apple eventually recognized this trend and decided to retire its music hardware products. Visit the following link for interesting facts and an assessment from Statista about why Apple said goodbye to music devices: `https://www.statista.com/chart/10469/apple-ipod-sales/`.

It's so convenient to just talk to a machine and ask it to play the song of your choice instantly. You don't have to move your hand, touch a button, or shuffle through a rack of CDs to find your favorite songs anymore. You can just ask the machine to play your choice of song and your virtual assistant device plays it. I call this the inflection point in the human history digitization journey. It opens up a world of voice commands, such as operating lights, refrigerators, heating, security cameras, and home service drones – it's truly an inflection point where machines and humans define how humans live, work, and play.

It's interesting to see trends around digital assistants such as Amazon Alexa. Feel free to read more about this at `https://safeatlast.co/blog/amazon-alexa-statistics/#gref`.

The last few years also heralded a shift in the way we communicate both personally and professionally; I remember growing up watching Star Trek, which became very popular during the 1970s. It had the concept of cellular phones, which became reality in the 1980s, first in Japan thanks to NTT. It was fiction coming true in just a few years, and today, we can reach anyone on the planet in just a few clicks with HD video quality. Today, more than 500 million meetings occur daily on Teams and Zoom combined, which is a staggering digital immersion of our lives in technology.

Technology's rapidly evolving adoption due to the pandemic is transforming industries, companies, and governments at a pace never seen before. The democratization of AI and the establishment of cloud technologies is giving birth to new ideas, companies, and risks that were never imagined before.

The pandemic disrupted education across the globe and affected millions of students. In response, education institutes implemented some forms of digital learning. Digital learning opened up new ways to learn independently of physical proximity between teachers and learners. Digital learning provides a new learning environment that has benefits and risks. Digital learning provides the convenience of attending classes from your home. It also provides an easy way for your friends to attend the same class, or anyone else to attend the class on your behalf. The vast majority of students had never attended online courses before the pandemic. The experience was equally new to teachers across all age brackets. While most students and teachers were busy adjusting to the new digital world, what went unnoticed was the risk of digital learning.

How the world eats has changed dramatically, thanks again to the pandemic. Just a decade back, ordering food mostly meant pizza. Nowadays, food delivery has become a global market worth more than $150 billion; it has more than tripled since 2017. Most food orders pre-pandemic were delivered by a driver employed by the restaurant. There were fewer payment methods, including cash on delivery. In the post-pandemic times, things have changed. Today, most customers order via their cell phones and through food delivery apps such as Uber Eats, Foodpanda, Zomato, and DoorDash. The food delivery business has its risks, such as the time it takes to deliver food, packing to maintain the food's quality, and theft of food while it's being delivered. Risks that go unnoticed are private information about what you eat, the time you order, and sensitive data such as credit card numbers that get transmitted and stored across multiple systems owned by various third parties in the delivery network.

Changes triggered by the pandemic were unexpected and fast. More important was the new world, which was more digital and stayed not just for a few days but for months across the globe during the pandemic. Changes impact our lives in different ways. Some of us embrace change faster than others. The digital habits induced by the pandemic are changing the way we learn things, make payments, order food, go shopping, and work.

For most of us, change brings uncertainty and loss of control. Digital changes are no different. Sometimes, changes in technology are inevitably agnostic to our liking or the rate at which we adopt them. Digital changes could include downloading an app from an app store if you want to purchase goods or services, which creates new digital habits. Digital changes such as "you must update the software or you will not be able to get new features again" evoke mixed responses. Attending calls on your favorite collaboration suite, such as Microsoft Teams or Zoom, and sharing your screen is a newly formed habit.

Changes around video calls and video meetings came in so fast that it's worthwhile looking at trends, as covered in the following links on the usage and statistics for leading video call providers, such as Microsoft Teams and Zoom during the pandemic years:

- *Zoom Revenue and Usage Statistics (2021) – Business of Apps*: `https://www.businessofapps.com/data/zoom-statistics/`
- *Microsoft Teams Revenue and Usage Statistics (2021) – Business of Apps*: `https://www.businessofapps.com/data/microsoft-teams-statistics/`

You may not like browsing the app stores offered by various phone or technology service providers, you may not like the new version of the operating system, or you may find sharing screens a very mundane activity; however, changes are inevitable.

Some changes require you to act, such as updating your phone's operating system, while some changes just soak into your life, such as browsing the internet or spending time on social networks, without any action needed from you. I call these changes *ambient* as they bring permanent changes to our lives. Moving from SMS to WhatsApp, Telegram, and Instagram are examples of ambient change. Driving a car to a new holiday location using digital maps is again an ambient change that has soaked into the lives of billions. Ambient changes come fast, without friction, with extremely low learning curves, and permanence. Ambient changes are what I am afraid of most. These changes bring in differentiated digital risk, giving humans almost no option to go back to the old ways of doing the same activity.

Well, don't lose your thoughts, and let me remind you what Hagrid said: *"No good sittin' worryin' abou' it. What's comin' will come, an' we'll meet it when it does."* What this phrase teaches us is to not worry and face the changes as and when they come.

In this chapter, we'll explore how rapidly we are getting fused into this digital web around us. This chapter also discusses the contentious fact that it is as though an invisible hand is guiding us to become absorbed in and addicted to this digital life, where risks are only visible toward the end of the journey. To begin with, we'll explore the following topics:

- **Digital transformation**: This covers the journey we have followed to get to the digital domain – how quickly the population at large is immersing itself into innovations, providing new ways of living and working, and the associated risks; yes, associated risks.
- **An invisible hand**: This covers how the invisible hand that is made up of convenience, ease, and gratification of having control, time-saving mechanisms, and an unprecedented level of access to services is pushing us into the digital life, and new kinds of digital and physical risks.

There is surely a digital tsunami ahead that has benefits, new experiences, and new risks.

Digital transformation

Computing has come a long way, and so has the use of computers. Computers have also morphed from the size of big rooms back in the 1980s to small chips in IoT devices.

Cars manufactured 40-50 years back were very, very different from the cars manufactured today. Modern cars are connected and have integrated maps. I used to find it difficult to park my car between two cars, but not any longer thanks to the parking assist features. Cruise control in cars today has been upgraded so that it's adaptive and the car can maintain its speed relative to cars around it. Cars today come with digital displays with touchscreens that show way more than just the speed and acceleration of the old days. Cars today come with cameras outside for parking assistance and traffic symbol alerts along the road, and inside for checking on driver drowsiness. Modern cars today can also run on their own using autonomous driving.

Technology has changed driving so much that you don't need humans today to even drive a car. The future of driving is without a human driver. The car industry has been transformed and continues on its transformation journey due to technology.

It's not just the car industry; any industry, including healthcare, manufacturing, music, and television, is transforming due to the use of technology.

As are you beginning to see on executive profiles on LinkedIn, digital transformation officers, chief digital officers, leaders in digital transformation, and digital transformation as a skill are becoming prominent.

So, what is digital transformation and what risks does it bring to our lives?

As computers have touched business processes and gained more intelligence in the form of what they can see, the ability to listen, operate a mechanical arm in car manufacture, or maneuver a car, humans began to realize new ways to use computers.

Business processes that were established in companies across the world have matured and are running effectively. Governments also have well-established processes such as the system of tax collection, and workers such as traffic cops ensure smooth traffic flow and issue tickets when they spot a speeding vehicle.

Digital technologies have changed our lives. Most of the time, we think that technological advancements are a thing of the future, which is not true. Society gives technology and technological advancements a sense of purpose, and innovations will continue to disrupt and change norms.

Why do we need a traffic cop when speeding cameras across the country can automatically issue tickets to drivers for speeding? Why do we need a human to process tax documents when AI software can assess, validate, and process the entire tax submission? Why do humans need to drive a car when cars can drive themselves?

These technological changes not only impact the normal way of working, studying, and socializing, but also impact actors, such as humans in the roles of customers, employees, partners, or other stakeholders that are part of the process.

Why do I need a human to deliver me pizza when a drone can deliver it faster to my doorstep by flying from a nearby pizza shop?

Let's look at an example of opening a bank account. We used to go to the bank with relevant documents, such as photo ID, proof of social status, and any other necessary documents. You would hand these documents to a bank officer, who would, in turn, verify them and open a bank account for you. This could take from a few minutes to a few days, depending on which bank and country you opened an account in. Once your account was open, you would receive a checkbook, a credit card, or a debit card to use with your account. Now, let's look at this same simple example through the lens of digital transformation, where you can open a bank account in minutes by using a mobile application, taking your picture, and entering your social security number, which gets verified by the government authorized agency such as the home or external affairs ministry online. Once it's verified, your credit rating is pulled from the credit agency and an account is created for you in minutes. A bank officer sipping their coffee miles away receives a notification on their phone to verify that the automatic account opening system should go ahead and create an account for this new customer in their banking system. The bank officer verifies your form, the picture you have taken from your phone, and your ratings from the credit agency, and presses the approve button.

In a matter of seconds, your phone screen says "thank you," displays your account number on your phone screen, and prompts you about whether you want a virtual debit or credit card instantly, while the bank sends you physical cards in due course. You are thrilled to get a new bank account in minutes and click on the virtual credit card, which again gets created in minutes for you.

Phew! That was the digital transformation of the account opening business process for a bank. It used a computer in the form of a mobile application that could fill in your details, take a photo of you from your mobile camera, and use a complex backend API and workflows that, in minutes, gave you a new bank account and a virtual credit card. You could be sitting in the Bahamas enjoying your coconut drink with the bank officer sitting thousands of miles away in a call center, facilitating your account opening without you even having to go to your nearest bank branch. It all looks great, but where does risk come in here? What if it was not you who requested an account, and someone else used your identity and photo to create a bank account and then misuses that account? What if in this process, you inserted a photo of Harry Potter, and the account got created without your image? What if, while creating a bank account, your personal information was also relayed and exfiltrated by an attacker for later use and abuse? What if your virtual credit card details got into the wrong hands? What if the wrong hands is not a human but software? What if this malicious software then makes a fraudulent transaction on your behalf? Who will you catch as there is no human in this process?

Digitization makes changes that use technology to make life easier for consumers, employees, businesses, and governments. It provides efficiency and new ways of achieving the same goals. It also creates new types of risk. Some of these risks are visible (known), while others are invisible. We'll explore this in more detail in the next section. Feel free to read the *Global trends: Navigating a world of disruption* report from McKinsey at `https://www.mckinsey.com/featured-insights/innovation-and-growth/navigating-a-world-of-disruption`.

An invisible hand

Today, we send more WhatsApp messages than we used to with SMSes just a few years back. Most of my friends don't send SMSes any longer. Nowadays, it's WhatsApp, Snapchat, or Telegram.

Do you know we also had a **multimedia messaging service** (**MMS**), which could be used to send images, video, or contacts to recipients? It was rich but it could not stand the onslaught of WhatsApp, which provided a much easier and more intuitive way to send and received images, videos, and audio messages.

So, what happened here? MMS capabilities used to be pre-installed on each phone, whereas users are required to install WhatsApp. Since both provide rich multimedia content, why did MMS not take off?

Let's look at user experience and touchpoints in using these technologies. SMS is off the table due to its limitations regarding handling multimedia content. Let's look at MMS and WhatsApp: both can send rich content, such as images and video, but MMS uses SMS transport as a channel to send content, so it will always require cellular phone connectivity with your provider. On the other hand, WhatsApp is independent of your cellular network. It works when you are on Wi-Fi or when you are without cellular connectivity.

To add to that, if you roam to other countries, sending MMSes can be very expensive as it also uses cellular infrastructure. What makes WhatsApp the king of messaging is that you can send messages even when there is no connectivity. Yes, the message gets in the sent queue and the user can do the next task without waiting for the message to be delivered to the recipient. A user can send the next message, leaving the WhatsApp framework to send messages as and when a connection is available. This is a very powerful feature; I call it *send it and forget it*. The verdict came quickly and MMS died at a very rapid speed, but what it left behind was the trace of an invisible hand at work. If the technology you produce is easy to use, cheap, and intuitive. it will get adopted very fast as if there is an invisible hand making it popular and adopted across masses, countries, and languages.

So, for a technology to truly offer an invisible hand, it needs to deliver greater efficiency and innovative capabilities that increase the value to customers multifold times. Primary capabilities must be easy to use (no manual, no tutorial, no how-to for using as many features as possible). It should be independent (that is, not tied to any platform, device, time, or channel), and have a free version. At the time of writing, digitization has increased and given all of us more thinking time to innovate; in essence, the pace of technological innovation has accelerated. Look at buzzing stock exchange companies that are more digital and tech-savvy; they are leading the pack and increasing their market share. Pick any company in insurance, finance, distribution, retail, telecom, or any other sector; the more tech-enabled the company is, the greater the chance that it will disrupt the market and become successful.

The next wave of innovation is led by AI, and its infusion will lead to more disruption and automation than any other technology on this planet. The following chart developed by McKinsey (the original graph can be found at `https://www.mckinsey.com/featured-insights/innovation-and-growth/navigating-a-world-of-disruption`) shows sectors and companies that will be impacted by AI:

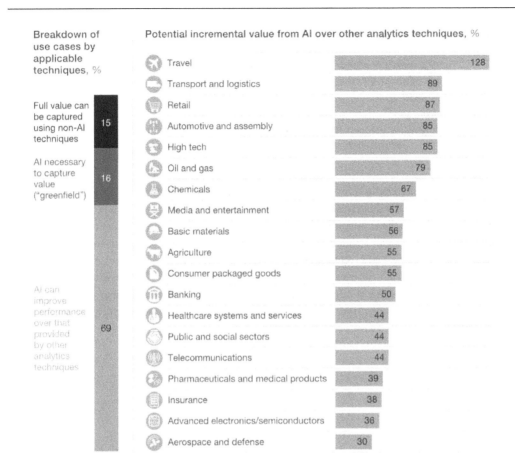

Figure 1.1 – The potential of AI to deliver more value across industry verticals

While the pace of adoption of AI by various companies across sectors will differ, the invisible hand that is driving digital adoption will be evident in the way employees work, engage with customers and partners, and collaborate with other employees. Business engagements will use more AI to reduce or ease human work. When we talk about AI, we always think of devices depicted in movies and science fiction novels.

AI tends to be accompanied by the thought that if everything is done by machines, robots, and AI, what will humans do? Some media companies and blog writers have exploited this, thinking with creative headings such as "the top 10 jobs that will be lost in the coming years." It's natural for people to read such articles to see whether and, if so, how their work will be affected. As per the McKinsey study, automation and AI promise to create more jobs than they will take away or replace redundant jobs and bring something new into the picture instead. Also, AI promises to make jobs easier while creating new jobs that will require different kinds of skills. For now, it is not a zero-sum game: AI promises to create more jobs.

Let's quickly look at the key message from the report:

> *"Under a midpoint scenario, about 15 percent of the global workforce, or the equivalent of about 400 million workers, could be displaced by automation from 2016 to 2030. At the same time, 550 million to 890 million new jobs could be created from productivity gains, innovation, and catalysts of new labor demand, including rising incomes in emerging economies and increased investment in infrastructure, real estate, energy, and technology."*
>
> *– McKinsey study 2021*

What also makes this interesting is my observation that 3 years in the IT industry is almost the same as 10 years in conventional industries due to the pace of innovation. While my observation's timeline is debatable, innovation in the IT sector led by Microsoft, Apple, Google, and Facebook is unparalleled in any other industry. The second point I want to make is the time it takes for an innovation to become mainstream is also reducing. I still recollect how quickly WhatsApp ate SMS for lunch and how Zoom/Teams ate telecoms voice calls for breakfast, or the time it took for consumers to switch from big fat TVs to thin smart TVs at home. While it's about adoption, the following visuals also share the speed at which we have adopted digitization. Let's take a look at the following graph (the original graph can be found at `https://www.visualcapitalist.com/rising-speed-technological-adoption/`) to understand this better:

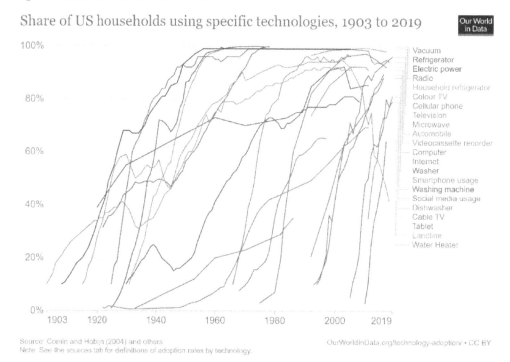

Figure 1.2 – Consumer rate of adoption for various technologies

Indeed, all of us in our various personas, from our working lives or personal lives, want to adopt and try out new technologies. I get goosebumps when young kids these days prefer to text with emoticons to express themselves because I find that such short and quick ways to display emotions are not the most efficient when it comes to human-to-human interaction. Our quick adoption of digitization is also creating new and different behavior patterns.

Summary

As we embrace digital life, all our activities, such as talking to a friend over a video call, planning a trip to our favorite destination, approving a purchase order created by our finance team member, having a board meeting or sensitive organization change discussion on a digital call, or communicating via email about a sensitive dividend or stock data with our banker, happens with a few clicks on our phone or computer. As I use my phone to check for the availability of hotels in Central London, I can surely feel two things strongly. First, the digital tsunami has just started and there is more to come. Second, the invisible hand of change will ensure populations across continents are immersed in this digital life, which will lead to more productivity, empowerment, and digital risks.

In the next chapter, we'll uncover more risks that technology delivers, along with the productivity benefits it provides in our lives.

2
Going Digital

As I sip my coffee, I check through my phone calendar for any upcoming meetings. I will be joining a Microsoft Teams meeting in seven minutes and am worried about the noise of drilling coming from the construction site nearby. I worry I may have to mute myself when not speaking on the call. But to my amazement, I discover that the Microsoft Teams software cancels out all background noises, including the sounds of drilling machines and hammering, saving me the stress, effort, and distraction of having to mute and unmute myself during the call.

What a breeze! Now, I start thinking that if intelligent software can cancel out noise, then could it also add noise such as the sound of traffic, an arriving train, or ambient airport sounds? I could fool someone into thinking that I am at the train station while I am taking a call from my home or a cafe. This can also extend to connecting with people online. Although I don't know the mobile numbers of my friends by heart, joining three of them on a call and talking to them is so easy. I don't have to call the first one, then call the second and third ones, and press the conference button on my phone today. I can just set up a single call and they can join it at their convenience. I feel like part of my life is missing if I don't have my phone for a few minutes.

Let's look at another scenario. My Teams call has started, and I feel hungry. I want to order something to eat, so I just grab my phone and point the camera at the barcode on the table. It's a contactless menu. I don't have to call the waiter to get the menu and then flip through the pages. The contactless menu gives me the names of the dishes, their pictures, nutritional details, videos and images of how the dishes are made, and some reviews. This is way more information than used to come on a standard menu. I can also quickly filter the choices by my favorite cuisine or by healthy options only. All of this started just with me pointing my camera at the contactless menu barcode, and my phone is still hosting the Teams call while I order my dish. But phones and small IoT devices can do a lot more than this to make our lives easier. Let's take a look at all of these changes happening around us in this chapter.

In this chapter, we'll cover the following topics:

- Hello Alexa, Siri, Google, Cortana, and more – new interfaces and intuitive means of human-machine interaction
- How software today is parsing so much data from our digital calendars that it can predict our actions in the physical world

- Cameras everywhere – am I being watched?

- Digital wearables – my watch is for much more than telling the time; it knows my schedule and health and even has my music and messages

- What is real?

Hello Alexa, Siri, Google, Cortana, and more

"We need a good user interface that's intuitive to use" – I keep hearing this phrase in corridor discussions on why it's important to have a good user interface. This topic has also given birth to design thinking, where software for the automobile industry is very focused on products and services having a good user experience. When I work with Microsoft Outlook, I am amazed at the way information is organized. The Outlook software that I use to check my emails also has my calendar and contacts. It helps keep my work life organized. Like most people, I change my laptop every few years, and it always strikes me how easy Outlook makes it to move my emails, contacts, and calendar information. It's the simplicity and intuitive user interface that makes me stick with Outlook for my email, contact, and collaboration needs.

If you're of the same generation as me, you may have purchased both the Microsoft Zune and the Apple iPod; these devices had a touchpad dial that let you control both the volume and song and provided a differentiated experience. A good and intuitive user interface keeps your users coming back to use the software. The touch material used in the device was soft and you could easily glide your fingers on it to increase or decrease the volume. It felt nice compared to pressing a button multiple times on any other music device to change the volume on the device. The user interface makes a difference – a big difference.

What Amazon did with Alexa in some ways took music devices to a whole new level as well. The designers asked the question, *why do you need to even touch a device for it to function? Why can't your device just listen to your voice commands and play music for you?* It's what I call out-of-the-box thinking and having the grit to experiment. It paid them rich dividends. Today, voice-based commands are present in other home assistants from Google, Apple, Microsoft, and other vendors, and millions of Alexa devices operate with voice commands. It's a new type of human-to-machine interaction – that is, without touch. As long as your voice can reach your device, it will play the music you ask for. *Figure 2.1* shows various user interface that can be operated via a click, touch, and voice:

Figure 2.1 – Different types of user interface – operated by a click, touch, or voice

What is important is how easily a user can complete their required interaction with a given machine. Sometimes, we measure this using the lowest number of clicks to perform the job, the amount of metadata that they see for a brief time, or the ease with which they can instruct their voice assistant to perform their operations.

A good user interface helps people use more software. It helps software become more popular with users. When more people use digital products, our personal and work lives become more digitized. Every piece of software and every digital device has a way to identify you. When you log in to a piece of software, it's called identity verification. For example, we all have an email ID and password that we use to log in to our email clients such as Gmail, Hotmail, or your company-provided email software. Every Amazon user has an identity (user ID and password) that allows them to log in to the Amazon website. Every iPhone is configured with a user identity. Once it's configured, this identity is used by software to recognize you as a user. This is how my phone knows to show me my calendar and emails. And just like my phone devices, I also have an Alexa device configured with my Amazon account credentials.

I must admit that I have many Alexa devices in my home. I find it very convenient, from switching on a specific lamp in my work room to all the lamps in my home in the evening – it's all just one call away. What I like most is the benefit of switching to a specific TV channel by voice in less than 2 seconds versus finding the remote, and then finding the channel button, switching to the main menu, and then taking a few more clicks to go to my desired channel. Being able to switch to a specific channel or set a specific AC temperature with just a voice command is what brings delight to customers in the smart device era.

Amazon Alexa also has a variety of partners who also build and supplement this automation functionality, adding to this huge ecosystem. We have global consumer giants who have built devices such as electric bulbs, geysers, microwaves, and electric switches and plugs that operate with Amazon Alexa voice commands. All these devices can be combined to create a smart home where electric devices can be operated using voice commands or programmed to switch on or off as scheduled. It's very nice to be in a house where all the lamps automatically switch on 15 minutes before sunset and are switched off automatically around midnight.

Being able to simply say, "*Alexa, keep my room temperature at 25 degrees Celsius and switch off the AC after 2 hours*" is truly a breeze. This will ensure that my ambient room temperature is kept to my liking and the AC automatically switches off after 2 hours, by which time I'll be asleep – and the list of things we can do with Alexa grows every day.

> **Note**
>
> You will find the following link handy to see the full extent of voice commands in the Amazon Alexa ecosystem: `https://www.cnet.com/home/smart-home/every-alexa-command-you-can-give-your-amazon-echo-smart-speaker-or-display/`.

From an authentication perspective, anyone who visits my place can also ask Amazon Alexa, for example, to play their favorite song. Yes, there is no authentication required for Alexa to play a song, or for that matter to switch on a light. Is that risky? Well, at least not in the case of playing songs for my friend who has come to visit.

Let's consider a few other scenarios. Today, in my home, both my air conditioner and water heater are voice-operated using the Amazon Alexa device. What if I asked Amazon Alexa to switch on the air conditioner and instead it switched on my geyser? Or if Alexa switched on my air conditioner while no one was at home? Every night, I issue a voice command to Alexa and say "*good night.*" The *good night* command is configured to switch off all the lights in my home. What if, even after I've given the voice command, Alexa is unable to send messages to the light bulbs to switch themselves off? In this case, I will have to go and physically power them off. What if there is a bug due to which Alexa is unable to issue the right command to the water heater, air conditioner, or bulbs to switch off? Or if Alexa turned on all my devices while I was away on vacation? That would lead to greater electricity consumption and a much higher electricity bill than expected. Besides malfunctioning, the possibility of digital assistants such as Alexa listening to my general conversations in addition to commands, analyzing my environment, and using this information without my consent creates privacy risks.

At the time of writing, Amazon has added more members to the Alexa family named Amazon View and Amazon Astro. Both of these products have the capability to move around your house. Yes, they can move, they can see, and they can talk. It's a new user interface of interaction.

The role of digital calendars

I spoke about my love for Outlook Calendar earlier. It's a nice interface that keeps me in control of what lies ahead in my day, such as the emails that I need to respond to and any meetings planned throughout the day.

I remember the old days of life without mobile phones when we used to have desk phones. Yes, we had emails, but mobile phones were still not widespread back then. I used to remember around 20 or 30 landline numbers of my friends, my home, and my office – key numbers to call in case I needed any help. Most of us were able to rattle them out in seconds from our memory.

Today, for most of us, it's difficult to remember the mobile number of more than two or three friends or family members. It's so easy to save a contact on your phone, just taking note of the person's name and letting the phone handle the mapping of name to number when you choose to call them. Contacts are a really simple feature on our phones, but very handy. I have stopped remembering the mobile numbers of my friends and family members altogether now.

Is there a risk involved in not knowing the phone numbers of people I know? What if I am in an emergency and my phone has run out of battery, has been stolen, or is just not functioning? How could I call my mom, dad, or best friend?

It's not just phone numbers, either – the same problem applies with the roads and routes we take when driving. Gone are the days when I used to remember directions such as "*go 30 miles east, then take a sharp right*" and look for road signs or key landmarks to validate my route. Now, I just use Google Maps or the mapping system built into my car to direct me, and reaching my destination is a breeze.

My reliance on Google Maps is so strong that I have actually even stopped looking at the built-in mapping system in my car. Moreover, nowadays car manufacturers apply technologies such as Android Auto so that their built-in mapping systems use real-time traffic data from Google. It's almost become the norm – in modern times, it's nearly impossible to drive around even in your own city without GPS. Again, it's the invisible hand that takes care of everything, and you become peripheral to the technology. Well, it's just a map – what could go wrong? In a worst-case scenario, it may lead you astray to the wrong location, and you'll have to drive back and see what alternative routes are available.

The real value Google Maps provides is real-time traffic updates, which are so useful for considering alternative routes we could take to reach our destinations in the shortest possible time. Guess what? Google gives you that information based on data it gathers in real time from other Google Maps users. So, could I trick Google into reporting a nonexistent traffic jam to its users? Hmm... is that possible?

Yes! An artist (although I'm not sure why he is called an artist) proved it by using just 99 phones, borrowing some from his friends:

- *Artist uses 99 phones to trick Google into traffic jam alert* (CNN Style): `https://edition.cnn.com/style/article/artist-google-traffic-jam-alert-trick-scli-intl/index.html`
- *Google Maps shuts down editing after the 'robot peeing' incident* (`cnn.com`): `https://money.cnn.com/2015/05/11/technology/google-maps-android-peeing/`

Imagine if there was just one connecting road between two cities and I did this stunt on a busy Monday morning, causing traffic to be re-routed. Also, on the new route, I happen to own an isolated cafe that now gets windfall profits. The risks of digitization are everywhere, as we will see in more detail in the upcoming chapters.

Most professionals today can't operate without a calendar. I manage both my personal and professional life using Outlook Calendar; it knows when I am free, when I am busy, and what I will be doing next week, keeping me aligned with my goals and helping me with time management. My phone queries

Google Maps for traffic en route and warns me that I may be late and miss my flight if I don't start moving now.

The phone also has access to my meeting agenda. Now, imagine if the subject line of the agenda contained a proposed dividend for a publicly listed company. What if the meeting notes in the agenda also proposed a specific percentage for the dividend for board approval, which makes perfect insider information? Well, thank God that the phone won't send this information to anyone else, but what if it could? Is it a risk to store such sensitive information on my phone? What if it gets stolen or left in a taxi?

As an example of the risk, imagine that I get an email and open it. It contains some hidden software that configures my phone to forward all my emails to some anonymous email address over the internet. The person behind this can then read all my sensitive, private information.

When information from my phone's email client is sent to another email address over the internet, it uses a **forward rule**. This is a simple rule that anyone with an email address can configure. Users might intentionally configure this rule to forward information to colleagues who could stand in for them while they are on vacation. The problem arises when people set rules and then forget to assess or validate them at regular intervals. Malware creators exploit this forgetful element of human nature and make their malware configure rules to forward emails to malicious email addresses, unbeknownst to the user. However, if you are using Microsoft O365 as your email platform, you can sit back and relax, as the email software blocks all forwarding rules by default, regardless of whether it is sending or receiving email or calendar information. The email application does this via **outbound rules** and can be used by attackers to steal information from our email systems. The following link provides more information on the role of outbound rules and how to configure outbound rules in a Microsoft email security system:

Configure outbound spam filtering - Office 365 | Microsoft Docs: `https://docs.microsoft.com/en-us/microsoft-365/security/office-365-security/configure-the-outbound-spam-policy?view=o365-worldwide`

Digital and physical socialization

In today's interconnected world, both the physical and digital realms can be sources of sensitive information that can reveal more than we might think. For instance, a seemingly innocuous Outlook Calendar can unwittingly give away confidential information; I discovered this when I stumbled upon a meeting with the subject line *"Resignation handoff for Bob."* Despite the information being known only to Bob's manager, the calendar revealed Bob's departure to me while I was simply trying to find an open meeting slot.

Similarly, in the physical world, interacting with strangers can be fraught with uncertainty, as I discovered when a stranger in Sydney began asking personal questions about my visit. Unsure of the stranger's motives, I felt pressured to share information that could potentially compromise my safety. He was asking seemingly simple travel-related questions – where I was from, when I arrived, how many days I was in Sydney, and about my hotel, workplace, and my family. We don't even realize that we reveal all the aforementioned information so easily in the public domain on social networks while posting pictures of our family or checking in at the airport or our favorite hotel.

In both cases, the need for caution and awareness is clear. Whether navigating the digital or physical world, it is important to be mindful of the information we share and the potential consequences of our actions. In the digital realm, this might mean being mindful of who has access to our online calendars or being cautious when clicking on suspicious links. In the physical world, it might mean being careful about who we interact with and what information we share, especially when we are alone or in an unfamiliar setting.

Of course, this doesn't mean that we must live in constant fear or paranoia. Rather, it is a reminder that our actions, both online and offline, can have unintended consequences, and that a little bit of caution can go a long way. By being aware of the potential risks and taking steps to mitigate them, we can better protect ourselves and our information, both in the digital and physical realms. If someone in the physical world asked for personal information, you may think, why should I tell them? Well, if you feel the same way, why do you give that same personal information to thousands of friends on Facebook, along with their friends, and many people you don't even know or care about in your friends list?

A simple check-in with a photo reveals to all your Facebook friends (depending on your settings) where you are at what time, whether you're alone or with someone, and what you're doing till when. There are different reasons that people broadcast check-ins to the world on Facebook, but no one wants this information to be misused. Sharing too much information on social media makes it easier for cybercriminals to target you in a phishing attack or misuse this information to gain access to social media accounts, answer security questions on financial sites, or send customized **spear-phishing** messages designed to fool you into handing over even more sensitive information. It's like an open invitation for attackers to hack into your digital accounts.

Simply by checking in somewhere on Facebook, you tell the world of strangers you are not at home, and if you are overseas, it's easy for a robber to estimate when you will return and plan a robbery of your home, assuming you live alone (which is also evident from your Facebook profile). What easy prey. The following is a great article that discusses the risks of checking in and updating your whereabouts on social media:

The Risks of 'Checking In' – Kaspersky Daily | Kaspersky official blog: `https://www.kaspersky.co.in/blog/risks-checking-in/1682/`

As we've seen, due to digitization, we are becoming more and more digital in our actions without realizing the potential risk it exposes us to.

Cameras everywhere

Most phones today have cameras with higher resolutions than the cameras available to professional photographers a few years back. Contemporary phone cameras offer features such as zoom and can even enhance photographs using automated software. But besides our phones, cameras are appearing in many other places. There are cameras in the office for security, and more and more people have followed this trend in their own homes. Residential communities, malls, and manufacturing facilities have cameras as part of security or productivity monitoring, or to meet regulatory requirements. Cameras on streets to capture traffic offenses and issue fines to offenders are increasingly the norm. It's a world of cameras everywhere, and it's quickly evolving – let's not forget that personal digital cameras appeared only 2 decades ago and were quickly phased out due to the rapid innovation in this space.

Before phones became our primary photo-taking devices, for a few years digital cameras invaded the market share of traditional film-based cameras. While Kodak was trying to protect its film-based market share, Sony unleashed a wave of innovative digital cameras. I remember the famous electronic markets in Singapore and Hong Kong were filled with digital camera shops, and buyers from all across the globe came to get their hands on the latest equipment.

The digitization tsunami changed the way users used cameras. First, users moved from large tape- and film-based cameras to digital cameras. For a user to see photos they had taken on a film camera, they had to give their film to a developer, pay them to process the film, and wait at least a few days.

The increased numbers of consumers using digital cameras disrupted the ecosystem of film production studios. When you take photos with a digital camera, you can see them instantly. Digital cameras' ease of use also increased the number of photos that users took, as there was no associated production cost to view the photo.

Sony, in particular, reaped the benefits of this wave of digital camera purchases. Newer versions of digital cameras came with better zoom functions and higher resolutions. Cameras in the year 2000 offered resolutions from 4 **megapixels** (**MP**) to 12 MP, with a 10x zoom feature. The digital camera boom continued from 2000 until the iPhone arrived.

In 2007, the iPhone was born. It was not the first phone to have a camera or a touchscreen, but it changed the phone market and became an inflection point in mobile technology. I remember there was no Apple store then; YouTube was still new (1–2 years in its infancy); I was still a Yahoo Search fan. Apple fans thronged in queues to get their hands on the iPhone, from stores in malls to the much-talked Apple Fifth Avenue in New York. The energy was the same everywhere.

The iPhone arrived with a built-in camera and provided an alternative option to users by offering a phone and digital camera in a single device. It was an easy choice – why would anyone buy a phone and a camera separately when they could get a phone with a built-in camera that provides the same (or better) resolution as a digital camera? As a result, phones started to become a unified and ubiquitous device, the focal point of all digital capabilities in our lives.

We will talk more about phones and phone-related risks later. For now, let's refocus on cameras.

Today, all laptops come with built-in cameras. Cameras are digital eyes that continuously observe the environment. Unlike human eyes, digital cameras don't blink or get tired and record everything that they see on computer memory, which can be replayed as required.

According to CNBC, there are close to 1 billion surveillance cameras alone globally. This count does not include phone cameras, home cameras, and laptop cameras:

One billion surveillance cameras will be watching globally in 2021, a new study says *(cnbc.com)*: `https://www.cnbc.com/2019/12/06/one-billion-surveillance-cameras-will-be-watching-globally-in-2021.html`

So, there are a billion eyes capturing the non-stop moving of goods, humans, and vehicles, and they allow the replay of things that happened in the past. This serves as a deterrent to criminals and helps solve cases by providing videos of what happened at a crime scene. Today, almost every law enforcement agency banks on evidence from cameras not even necessarily owned by them to solve cases. Cameras are not only here to stay but will also grow in number and become more intelligent.

This "intelligence" comes from connecting camera feeds to software that uses artificial intelligence algorithms to detect what's visible to the camera and then create alerts or actions determined by human users.

I remember just a few months back, while driving on a highway to Delhi, I was caught speeding by a camera. I got an email with a picture of me taken by this smart camera, detailing the fine that I had to pay. Likely, the software behind the camera read the numberplate of my car from the live camera feed, and the camera was also fitted with a speed monitor to assess the speed of my vehicle. Once the software determined that my speed was more than permitted, it then made me a candidate for a fine. The software might have even queried a database to find out the car owner's details and then sent me an email. I am assuming the workflow here – any human operator would have either accepted the guidance from the software or maybe just let the software decide itself in clear-cut cases. In either case, we are being watched 24/7 by digital eyes with the power to take action.

We spoke about Google Maps previously. Recently, Google released sets of images of certain areas from various satellites over time, combining them into timelapses to show how a particular area changed in the last few decades. You can learn more about this by visiting the following web page: *Earth Timelapse video downloads – Google Earth Engine*: `https://developers.google.com/earth-engine/timelapse/videos`.

The cameras on satellites, especially military satellites, are a closely guarded secret. An article by Forbes detailed that most commercial, non-military satellite operators have resolutions up to only 25 cm. Conversely, one of these satellites can identify a $100 note clearly from 400+ km up in the sky.

The element of risk here arises when cameras start sharing, leaking, or using data in ways other than the primary use case. For example, let's say I go to the ATM to withdraw cash but an attacker has hacked into the bank's security camera to pick up my PIN. What if the cameras you had in your boardroom could be operated at will by an intruder to eavesdrop on your board's discussions? What if someone was listening throughout your laptop as well?

The next example is not only scary but also operates in a huge commercial market. It's branded as a tool for modern parents to track their children, with subscription plans starting from $49. It's the market of *mobile spyware* that works in stealth mode, and at a click of a button by a parent (I prefer calling them the "handler of spyware"), it shares pictures, voice call data, and a child's current location, and they can also switch on a phone's mic to listen in to what's going on and even discretely operate the camera. It's not privacy invasion – privacy isn't in the picture at all. There's a huge market for this, with the top 10 service providers ranked by the quality of service they offer.

We are fully immersed in digital life, but the risks, as we've seen, include tracking systems invisible to the human eye. I will suggest you visit `https://www.mspy.com/` and see for yourself how these spy apps are marketed and can be purchased by anyone on the internet.

Digital wearables – oh, my heart

The wave of digitization is huge and continues to extend into various new use cases. One new area that is growing in popularity is digital smartwatches and health-related IoT devices. The most common form factor remains smartwatches, but there are other niche offerings in the health sector, such as smart rings and new types of devices that can track your health stats just by reading signals off your skin. The most popular that comes to mind is the Abbott glucose metering smart device:

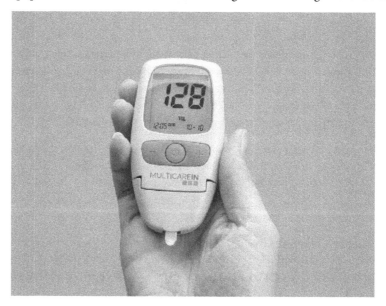

Figure 2.2 – Digital wearables, that is, wearables that stick to the skin for months

The preceding photo can be found at `abbott.com`.

Digital watches have been on the market for some time, but as with phones, the big changes came from Apple delivering precision and expanded features in its watches. It was unthinkable a few years back that a watch on your wrist could tell you your blood oxygen level, be able to take ECGs, or monitor your sleep time. Apple has positioned its watches as the ultimate device for a healthy life, creating an equivalence relation between a good life, its brand, and health.

The COVID-19 pandemic brought an immense focus on healthcare infrastructure across the globe. Most of us didn't think we would be using terms such as *quarantine* on a regular basis and considering the importance it holds in breaking the chain of infection. Most of us learned more about vaccines, the time it takes to develop them, the process of testing from clinical trials to human trials, technologies such as mRNA versus viral vector methods, the difference between *endemic* and *pandemic*, and the impact that both have on the economy and social life.

In the IT world, we have been using the quarantine process for over a decade for machines that get infected with computer viruses, and it parallels what we practice in the physical world. In the physical world, we ask a person to isolate at home; in the digital world, we ensure that an infected machine is isolated in a newly created network and no other computer can communicate with the machine.

Today, wearable devices are becoming more intelligent and come in various forms and sizes. Devices can detect your vital health parameters, the environmental temperature, the humidity, pollution levels, light variations, and noise levels, and either take action on their own or give the user of the device a prompt to approve the suggested action. Most wearable devices also pair up with our phones, using the phone's ability to connect with the internet and access intelligent cloud services.

A device can monitor your heart rate, and if it detects an abnormality, it gives you a prompt on your phone or reaches out to a web service on the internet to see whether it needs to call your emergency contact. What if a wearable device detects a heart rate that is highly abnormal, calls your doctor, and it's found to be a false positive? How much should I trust the ECG report from my Apple Watch? Likewise, what if I was experiencing genuine symptoms of a heart ailment but my device didn't detect it – a false negative? Is this a risk, or just the inability of the device to accurately take readings and, hence, produce a false or delayed alert?

IDC is a research firm that tracks the worldwide wearable device market, which includes devices such as smartwatches, fitness trackers, and ear-worn devices. The top 5 vendors in terms of market share are depicted in the following figure, based on IDC's latest report (Q12021):

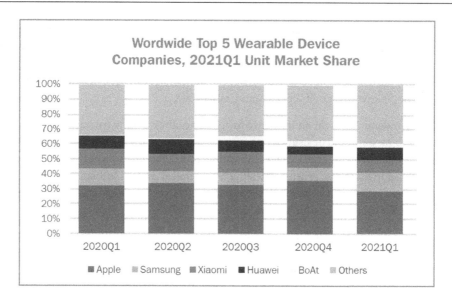

Figure 2.3 – Worldwide trends for wearable devices

Overall, the wearables market continues to grow, with shipments reaching 45.1 million units in Q3 2021, up from 34.3 million units in the same quarter last year. Smartwatches continue to be the most popular category of wearable device, accounting for over half of all shipments in the smartwatch category, with Apple and Samsung leading the pack. Shipments of Apple watches have again started topping 100 million a quarter, back to pre-pandemic levels. The specs of recently released digital watches from Apple, Samsung, and other leading vendors promise improvement and enhancements of more health tracking features, such as heart rate, in addition to oxygen level monitoring and the ECG app that were available in previous versions.

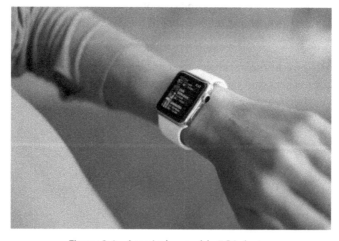

Figure 2.4 – A typical wearable ECG device

According to information on the Apple website, it's easy to guess what to expect in the upcoming version. Currently, the watch is built with sensors that when placed against the wrist can read health signals such as blood pressure, heart rate, and the oxygen level in the body. By adding more sensors, the watch will be able to detect more and more health and environmental signals.

Advanced features in wearables

Many users enjoy the easy-to-find features in wearables such as the Apple Watch. These wearables may even replace the ecosystem of health-testing labs, much as digital cameras did with the photo development business. The Apple Watch use electrodes built into the Digital Crown that work together with the ECG app to read your heart's electrical signals. Simply touch the Digital Crown to generate an ECG waveform in just 30 seconds.

The ECG app can indicate whether your heart rhythm shows signs of atrial fibrillation—a serious form of irregular heart rhythm—or sinus rhythm, which means your heart is beating in a normal pattern, giving you a near-accurate ECG report.

In the absence of a watch-generated ECG, I would have to go to a nearby hospital or a medical device-equipped clinic. These wearables, therefore, form another inflection point powered by digital innovation.

It reminds me of my childhood days when the doctor checked my wrist and asked whether I could feel pain from him pressing my palm, as well as checking the dilation of my pupils, and prescribing me medicine. It worked like magic, and I was intrigued by how the doctor could tell whether I was okay or not with just the touch of a hand. This ancient science is now being reinterpreted by electrodes in digital devices as part of the digital tsunami that is covering all aspects of our lives.

What is real?

Today, we navigate the internet using browsers. We can search by just typing a brand name into the browser to reach its website. Typically, a web address is made up of a company name, followed by .com or a relevant country domain.

Website browsing means navigating to relevant web pages after you have opened a company's main landing page, and seeking the relevant pages, products, or information you are looking for by making a few clicks on the main landing page.

You must know what you are looking for and must look in the right places to find that information. You need to use your eyes and brain to spot the section and pages on the home page of the website. Once you go to the right section, you need to scan the entire page to see whether it has what you are looking for. If so, then you can click further and follow the rabbit hole that it leads you down. It's very simple but that's how the internet taught us to find information.

What if there was another way to find the information you were looking for – do you really need a website? Why should you scan sections of pages searching for info? Why can't you just ask someone digitally, and get the required response? This leads us to the metaverse. What exactly is the metaverse?

Right now, those involved in its creation cannot agree on any one definition. I think the metaverse is best characterized as an evolution of today's internet – it is something we are fully immersed in instead of browsing text-based websites. It may realize the promise of vast digital worlds that parallel our physical world, impacting how we shop and bank and making these things as immersive as playing video games.

Let's say I am looking to open a stock trading account in Singapore. I will visit my favorite stock broker website (let's say I prefer DBS Vickers (`https://www.dbsvickers.com/vickers/accounts/default.page`) as my broker, based in Singapore). My eyes will scan the site in seconds, and my brain will tell me to look for words such as **account**, **new user**, **open account**, and similar phrases. Luckily, I find **accounts** and then open a new account in the third tab. However, my eyes are continuously parsing text to find this information, and I have to click four or five times to arrive at this point.

Now, let's consider how a website full of text and clicking segments could be transformed into a more human-like experience that the industry refers to as a **digital human** for browsing and finding information. The following figure features the digital human Sophie from the company `digitalhumans.com`:

Figure 2.5 – User interfaces from click-based browsing to voice-based talking

What if the website of DBS Vickers consisted only of a digital persona such as Sophie, as seen in the previous figure? Sophie could converse with users visiting the website instead of them having to click around to find information. What if a potential visitor to the website could just ask Sophie to open an account, much as they would in the physical office of DBS Vickers?

I am sure you will find it fascinating talking to Sophie by visiting `https://sophie.digitalhumans.com/`. She is designed to be spoken aloud to, but if a user is in a public place and doesn't want to use voice as a medium for communication, they can instead use the keyboard to chat.

Summary

Humans, corporations, and even digital devices are in a race to become more digital. Phones become more than phones, watches become more than watches, a healthcare company wants to become a health-tech sector leader, a washing machine manufacturer wants to become an automation leader, and a car manufacturer wants to become a tech leader in the transport space.

Every digital technology has its own risks. Digital systems are so embedded in daily life that the potential damage from even a single security incident is magnified. In the next chapter, we will examine the types of risk that this creates and how to overcome them and be even more productive.

Bon digital voyage!

3

Visible and Invisible Risks

Sometimes, my mind ponders over how digital my life has become, and I wonder whether it is even possible to live a meaningful life without a phone. My phone structures my life – from meetings, music, and connections with my family and friends to office work, emails, news, stocks, travel, maps, and a lot more.

It's tough to have a productive day without a smartphone – it assists you in traveling (Google Maps/ Uber), meetings (Outlook and Teams/Zoom, among others), keeping in touch with family and friends (WhatsApp/Signal/Telegram), and ordering food when you are back home. As I sink into my chair to check my emails for the day, I tend to think about what kind of risks I am inviting by having so much of my work (and life in general) depend upon my phone.

What is a risk? Let's try and look at some scenarios to help us understand that. What if someone has cloned my mobile SIM card and can get my banking **One-Time Password** (**OTP**) for transactions, or financial transaction information? Today, my browser knows what I view, what I search for, how much time I spend on which site, what my preferences are, what I like to read, and what I like in general. My phone can read every one of my personal messages. The data on my phone (including personal messages, SMS, call records, and photos) are all stored and synced to the phone vendor's cloud. What are the risks if all this personal information is used by advertisers or gets accessed by cybercriminals?

Is everyone on the planet who uses a computer, phone, or tablet at risk? Can I do something about it? What if something malicious happens – what will that look like? What about companies that provide us with phones, computers, and all these digital technologies – what are they doing to protect us from risks?

Innovating new products and services that can help humans and companies to be more productive is the mantra, mission, and focus of lots of big companies. Technically speaking, cloud computing, artificial intelligence, IoT, data lakes, data analytics, robotics, digital humans, and bots will bring more digital elements to our lives. What are the true risks of using these technologies?

Companies and individuals must understand the risk exposure these technologies bring into our lives. This chapter focuses on risks – both visible and invisible – in digital life. As more and more services become digital, this chapter will help you identify, explore, evaluate, and mitigate digital and technological risks for a happier and more fulfilling life.

In this chapter, we will cover the following topics:

- Risks in digital life

- Visible risks

- Invisible risks

- When does risk become visible?

Risks in digital life

What are the risks when I use my phone or computer that is connected to the internet, use Amazon Alexa-operated lights and music devices, or internet-connected TV or cars?

What is *risk*? In simple terms, I would define it as the *possibility of something bad happening*. This brings to light two elements:

- What is the definition of *something bad*? Is something bad or undesirable for you also undesirable for everyone else?

- The use of the word *possibility*. It brings in an element of chance, an element of whether it may or may not happen.

Most of us think of chance, possibility, and probability as synonyms. However, there is a critical difference between them, and a lot of the time, we tend to ignore the big, thick line between *possibility* and *probability*. Let me explain it to you with simple examples from our daily lives.

Millions of people travel in modern aircraft every day. Lots of them are still afraid of flying, especially when the flight hits air turbulence. When in flight, the human brain thinks of two outcomes – first, the flight will land safely, and second, the flight may not land safely and something bad may happen. So, we have two possibilities – either the flight will land safely, or it will not. People all over the world take flights. Data from various sources (including the United States Department of Transportation and third-party sources such as FlyFright) shows that traveling by plane is much safer than traveling via cars or other modes of transport. So, the difference lies in the *probability* of something bad happening, not the *possibility*.

You can look at the data that's available publicly on flight safety by visiting the following link: `https://www.bts.gov/content/us-air-carrier-safety-data`.

The *possibility* of a plane crash (or a fan falling on you, for example) is 50%, but if we talk about the *probability* of it happening, it's probably a 1-in-3.37 billion chance based on data collected by `flyfright.com` from flight records in the United States for 2012-2016. Most business leaders leverage possibility and probability to their advantage and see the dark line that lies between them. Let's return to our case by defining what risk is.

Just to bring our attention back to the definition shared earlier and expand it slightly – *risk is the probability of an event occurring that leads to a negative or undesirable impact.*

Now that we understand the difference between possibility and probability in terms of assessing risk, let's apply that to our fan example. So, the possibility of a plane crash is 50% – that is, it will either crash or land safely. The probability of the same event is <.0001% based on the computation from `flyfright.com`.

Falling fans and traveling in airplanes both have risks associated with them. Falling fans can cause injury or death if they fall from a significant height and hit someone, while traveling in airplanes has risks such as turbulence, mechanical failure, and the possibility of a crash. It is important to understand that the possibility of both a fan and an airplane falling is the same, but the probability of both cases is extremely low. That's the reason billions of people all over the world take flights for business and pleasure.

Let's move from the physical world to the digital world and start applying the concepts of possibility and probability. Let's assess my digital devices with a possibility and probability lens in the context of digital risks that my phone brings with its usage. When I use my device, my phone, or my laptop, I am certain that it has been in my possession since purchase. I have high confidence that my device is not infected with malware that will steal my password. (This malware is also called a keyboard logger, and it will collect anything that you type, including usernames and passwords.) When I am invited to log in to any software email, WhatsApp web, or social media account from someone else's laptop or a cyber café machine, I am hesitant. I am at risk of my password being stolen, cached, or snooped. In this case, the probability of a machine being infected is very high, and as a cautious user, I would refrain from using an unknown device.

The story is incomplete when we talk about risk, and we leave risk mitigation out of the discussion. So, what is risk mitigation exactly? Risk mitigation talks about what you should do if you have a visible risk to prevent that risk from impacting you. Broadly speaking, risk mitigation falls under the topic of risk management, which involves identifying, evaluating, and prioritizing risks. ISO 31000 is one of the leading global standards that covers risk management and best practices. You can read more about it here: `https://www.iso.org/iso-31000-risk-management.html`.

ISO 31000 highlights activities, actions, and configurations that we can pursue to mitigate risks that have a high probability of occurring and could impact our goals significantly. In my experience, it's easy to put a probability on risks that I am aware of. What about the risks that I am not aware of, risks that are not visible today, risks that have never occurred or been recorded in the past? Let me start by broadly dividing risks into known risks (or visible risks) and unknown risks (or invisible risks). We'll look at them while covering a few examples in the next section.

Visible risks

To understand visible risks, we will look at some scenario-based examples. Let's suppose I can steal your password – let's say your social network password, be it Facebook, Twitter, Instagram, or whatever you use. That is a risk as it'll give me access to your private information. Now, what if I post something that puts you in trouble legally? What if I post something using your account against the government, the company where you work, or something that enrages society or your friends? Is that risky? Of course! It can have very serious consequences too.

Now, let me share the possible ways in which I can steal your social network password. Broadly, to do so, I will have to choose between hacking Facebook servers or hacking your system. It's easy to assume that hacking into Facebook servers is far more difficult than hacking into your phone or PC to steal your password.

Let's say I choose to steal your password from your mobile/PC. What options do I have? Do I need to meet you and have physical access to your phone? Well, that might be difficult – what if I don't know you? That makes my physical access steal method almost impossible. Again, I have multiple methods to think of, including remotely entering your operating system (either Windows, iOS, or Android, depending on what you use) and then waiting for you to enter your password and stealing it via keyboard logging or screen capture methods. It sounds very techie, nerdy, and complex. What if I had a simpler way to steal your password? Well, there is a simpler way – a technique called **phishing**. In common terms, phishing involves stealing a user's password by baiting them with a fake invitation to log in to some software.

In this technique, I need to know your email address, which is easy to get. I will then send you an email that looks like it's coming from your social network, prompting you to enter your password. But hang on – why would you need to enter your password? Well, I will craft the email in such a way that it lures you to enter your password as a way to verify that you are truly you. Let's say that I could create a fear that your password is expiring, and that you need to change it today. I could also say that based on your recent review on Facebook, you have won something such as a $10 voucher, which you can claim by logging into Facebook. As the page is set up by me playing the role of the attacker, once you enter the password on my landing page, which looks like a Facebook page, I will steal or phish your password. Now that I have your password, I can use your account as I please.

Phishing is still one of the most popular techniques used by attackers to steal passwords and present various risks to users on their digital assets, from social networks and e-commerce websites to corporate credentials.

The pandemic made this technique more effective, as more people today rely on email communication. Also, a curiosity to open emails that contain pandemic-related information makes for good bait for password stealers.

When identifying a phishing email, the attackers try to make the email look like a legitimate email by copying logos, text, fonts, and the placement of the email while following the pattern of that social network communication. There are ways for smart, well-versed users to identify phishing emails though. Being phished is a visible and present risk.

Let's talk about another email-based technique used by attackers. Mostly, this technique is related to the lottery, or some payout. Let me share a similar-looking email that I received a few weeks back. It was an email from Helene Bassett, as shown in the following figure, with a very alluring subject line stating **Urgent reply**. It asked me to contact her as part of her husband wishing to get financial benefits:

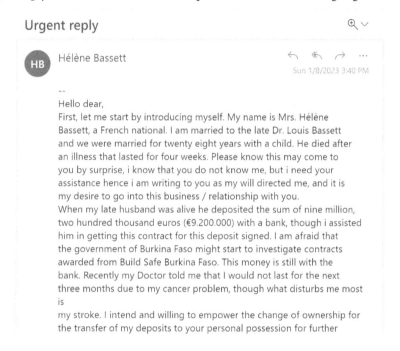

Figure 3.1 – Email phishing example

In some phishing cases, cybercriminals even make the communication look authentic by using rubber stamps and letterhead attachments with brand logos:

BRITISH CFP - UNCLAIMED FUND GIVE AWAY & CFP UNIT.

No: 124-134 Corbyn Street,
Islington, London N4 3DE,
United Kingdom.

Re: 2019 Unclaimed Fund Give Away.

Attention: Beneficiary.

Welcome to the 2019 World Bank/Paris Club & British CFP Global Financial SET-UP. The Camelot National Lottery Group has noted the wide spread of fraud activities that are being committed by charlatans & impostures within United Kingdom, Africa & the World in general. Those fraudsters use names of the United Kingdom International Lottery Companies to carryout their illicit act. Because of this reason, the World Bank/Paris Club & British CFP in collaboration with the Camelot National Lottery has set up a welfare program to compensate all those that have fallen victims and lost their money in transactions of this nature within UK, Africa & World wide.

In our quest, effort and effectiveness on Financial SET-UP. We have focused and committed to investigate all financial transactions going on all around UK & all over the world. Our Team of Google System Intensive Monetary Researchers [TGSIMR] had focused properly on every uncompleted local & international financial transaction transactions. Note: This Innovation was introduced by the CAMELOT NATIONAL LOTTERY in collaboration with the World Bank/Paris Club & British Government to help in completing every abandoned financial transaction to prevent re-emergence recession across the World.

In conclusion of our last Local/International Board forum which subject was "INVESTIGATION ON UNCOMPLETED FUND TRANSFER" Our Team of Google System Intensive Monetary Researchers [TGSIMR] after going through your last uncompleted financial transaction records with one of the commercial Banks, we have ascertained that you are one of scam victims and you have been approved to receive a compensation fund of £5,000,000.00. Therefore, the British CFP in collaboration with the CAMELOT NATIONAL LOTTERY is hereby issuing you a compensation payment prize £5,000,000.00 [Five Million British Pounds Sterling's] following the reveals made on your last abandoned fund transfer.

This compensation fund give away is done annually here in UK and funds are being generated from Unclaimed Lottery Fund from CAMELOT NATIONAL LOTTERY COMPANY. These unclaimed winning funds are either those winners' were automatically disqualified due to their age or from winners that turned up to claim their funds in our recent sweepstakes. Your compensation fund Winning Numbers are: XVX189298. For you to confirm that this letter is from Camelot National Lottery Group. Visit: http://www.national-lottery.co.uk/player/p/results/unclaimedPrizes.do

An arrangement has been made to pay your funds through (ATM Credit Card or Banker's Draft) to enable you receive your compensation fund without any interference.

Re-confirm your details to us as shown below for immediate action.

1) Full Name:
2) Physical Address where the ATM Card or Draft will be directed:
3) Telephone number:
4) Mobile Number:
5) Occupation:
6) Age:

Send your details as requested above on this e-mail: patrickwilson@execs.com: Name: Mr. Patrick Wilson, Phone: +27 (0) 63 598 4897 to confirm that you have received this notification and we shall get back to you within 24hrs.

Faithfully,

Mr. Hendric Gyan
British CFP Office.

APPROVED

Directors & Members.

Figure 3.2 – Phishing email contents

Another phishing email I received had an attachment (a letter) and it came with no text other than an alluring subject line. On reading the attached document, it looked like an authentic, stamped, approved document with links to the organization that seemed to be legitimate. It told me that I had been granted a few thousand pounds, and Dr Patrick has requested me to send my personal details, including my bank details. Well, I know that it's a trap, but millions of people who are not aware of this technique might fall for it. Is there any risk in sharing my bank details? I don't think so. There is no risk in sharing a bank account number, but what will happen after you send your bank details is you get an email back, saying that they can't send you the amount as your bank details are not complete. Let me share one handy technique. The email sender, after talking to you, may ask you to verify yourself by making a small transaction on their website to prove your identity. The website link they give will help them receive all critical information, such as your credit card number, secure code number (also called CVV), expiry date on the card, and other personal information, including your birth date and social security number. Another common technique is to lure you into donating to a website set up by them, showing images of underprivileged children. We also have more creative ones involving adoption, charity, vehicle sales, reverse mortgage, and others.

Feel free to browse `https://www.fbi.gov/scams-and-safety/common-scams-and-crimes` to keep yourself updated. I still call them visible risks, as most of us can identify these risks with some practice and awareness.

Next on my list are frauds committed by luring people in with great discount offers. It's the perfect trap for a shopping enthusiast. I must admit, I also once got caught in this trap and only realized it after I had paid the money. Let me share this true story with you. As I browsed Facebook, I came across an advertisement to purchase a car accessory. It was 70% off, as the advertisement claimed that this company was closing and selling all its inventory at a cheap price. On clicking on the advertisement, it took me to the website of this company. It was a typical e-commerce portal, with thousands of products, and almost all of them were at a huge discount. The site also communicated on its main landing page that there were 3 days until closure, so hurry and purchase while stocks last.

I went ahead, added the car accessory I had in mind, and paid around $10. Just after paying, I thought, *is this for real?* What caught my attention was the message after payment, which read "*We are experiencing a high number of orders and it may take up to 2 weeks to ship your orders.*" Guess what – I never received the product. I immediately went to contact them on their website and saw a support address with a Gmail domain. I was sure I had been conned.

I did put in a complaint with the cyber police and was sure that by the time my complaint would be investigated, this company would have vanished. That is exactly what happened. The lesson was to not fall for great discounts on websites you can't trust as it'll be a form of fake web shopping site fraud. I still put this in the category of visible risks.

Job frauds are even harder to predict. Job fraud refers to fraudulent or deceptive practices used by individuals or companies in the job hiring process. This can include posting fake job listings, misrepresenting job duties or requirements, collecting personal information or money from job seekers, or offering non-existent jobs. Some common signs of job fraud include being asked for personal information or money upfront, being offered a job without an interview, or being asked to pay for training or equipment. It's important for job seekers to be vigilant and thoroughly research any job opportunity before providing personal information or money, and to be wary of any job offer that seems too good to be true. Attackers exploit this human desire to grow and they will befriend people by pretending to be recruiters. They even set up a call to understand more about you and your profile, just like a corporate process of assessing candidates. As part of the interview, they will get your personal details, not just limited to your date of birth and your salary, as well as incentive details. Common personal details include your social security number, driver's license, your spouse's name, and any previous address that you have stayed at. On the pretext of verifying the same, they may also request your bank details or a letter from your bank showing details about your compensation or salary details. To make it more realistic, they may also get you interviewed for an overseas job position with their gang members based abroad, and toward the end, they will also give you an offer letter but ask you to deposit some amount toward visa or relocation fees. This amount will run into a few thousand dollars and once you deposit the amount, you will find that they are unreachable – their mobile phone will be deactivated, and you will get an email saying they're out of the office on a vacation for 2 weeks. After those 2 weeks, no one is traceable. Welcome again to the world of digital risks. It's still a visible risk. I believe that the attackers use hundreds of mobile SIM cards and keep rotating them to operate under the radar. Mostly, these SIM cards are procured on fake IDs, and they use virtual calling services with stolen credit cards to avoid any traces and detection.

Another risk that is worth noticing and ensuring that we protect our kids, young and old, from is sextortion. In this digital crime, the attacker can be a known or unknown person, and in most cases, uses an anonymous or fake identity. In this technique, the attacker will entice and manipulate the victim to share their private and nude photographs or videos over the phone, which are then stored and saved by the attacker for future extortion (sextortion). To begin with, the attacker will gain the trust of the victim and build an emotional bond with them. The attacker will share their own private photos to further cement that bond. Once this bond has been cemented, the attacker will exploit it by asking the victim to share their private photos and will keep collecting them until they have enough to use for extortion.

Another variation of this is to connect with male profiles over popular mediums such as Instagram or dating apps. Here, attackers use a fake profile pretending to be an attractive woman. The attacker will elevate the discussion into a sex chat and then ask for a nude call. When the victim agrees to it, they record the entire episode and start extorting the victim for money by threatening to share the video with his relatives and friends.

In a variant focusing on young kids, the attacker will forge a relationship with a young victim online and will then arrange to meet the child to exchange books and gifts. Later, the relationship turns ugly as they start using and abusing the child. In some extreme cases, the attacker takes the child to lonely places and coerces the child into producing sexually explicit images or videos through manipulation, gifts, or threats. All of this comes under sextortion.

All of us (as parents, friends, and colleagues) need to be aware of this technique. Especially for parents, it is important to show that they are aware of this and provide an open environment for their kids to report any such engagement in their circles.

The following are some links that you can visit to become more aware of such practices and new modus operandi used by cyber attackers:

- `https://www.fbi.gov/scams-and-safety`
- `https://www.nationalcrimeagency.gov.uk/`
- `http://www.cybercelldelhi.in/`

Congratulations! You are now more aware of the risks than you may have had little to no idea about before reading this chapter. All these risks are visible, and almost all of these risks have a technique and pattern. Most of these risk techniques have been used by attackers in the past in traditional or new ways. In the next section, we'll make ourselves aware of risks that blend so well into our digital life that even a cautious user is not able to spot them. I call these invisible risks.

Invisible risks

What if a cautious and aware user can identify a phishing email, or a user tries to check the legitimacy of a website before making a shopping purchase? What if the attackers are tasked to phish even alert and cautious users? What if the attackers are spending more time ensuring they are not exhibiting any known visible marks so that they aren't detected by highly trained and cautious users? What if the attack techniques are such that they only become visible over time and by then, users have fallen prey to the attack? This happens when bad actors perform an activity that involves using technology to deceive or cheat individuals or organizations and exploiting innocent victims by using their personal information or trust to commit fraud. This can take many forms, such as phishing scams, identity theft, or impersonating a legitimate institution or individual. Individuals need to be aware of these types of invisible risks. Let's look at a few more scenarios of invisible risk.

Regardless of the type of risk, visible or invisible, it can put the user in financial, physical, or mental/emotional distress.

Let's start with what I call **drive-by risks**. I call them *drive-by* as the victim's device can get infected by merely browsing a website. Most of us who spend time browsing the internet consider it to be a safe and normal way of finding information. Typically, when browsing, we start by going to well-known websites or searching on search engines such as Google and Bing. Let's say I am searching for a good café near my house or a good gardening service for my lawn in my neighborhood. The search result will show hundreds of sites that I don't know about. I click on what is shown to me on the first page. A click could take you to the website of that service provider – in my case, a lawn mower's website:

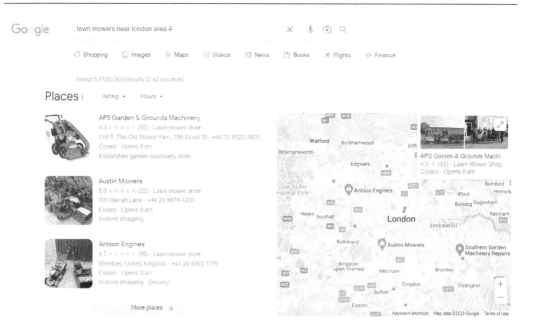

Figure 3.3 – Search engine redirecting to websites of lawnmowers in London

Through Google search results, you will land on a website that you may have not visited otherwise:

Figure 3.4 – One of many lawnmower service providers in London

This is a familiar page that millions of users get once they type their search query in any search engine, such as Google or Bing. You will get a list of websites relevant to your search keyword. What comes at the top of the list are the most relevant websites, as per search provider intelligence. So, in my example, if am searching for lawnmowers in London, Google search results would show me a list of service providers and their websites. I may or may not know anyone displayed in the search results. What if the website we clicked on contained malware that gets downloaded to your computer because of browsing? Malware comes in many forms, including information stealing, installing unauthorized software, privacy information stealing, lethal ones that can corrupt or delete all information in the computer, or asking for money (termed as ransomware). Malware can slow your machine down because of using your computer resources for other activities such as crypto mining. Malware can install keyboard logger software that can track every button you click and everything you type on your machine and send it to its master. Once keyboard loggers send all information that you type (including your messages, usernames, and passwords), it's easy for the master of this malware to then steal your credentials or be aware of everything that you type and operate on your devices. Malware can also install a backdoor on your device so that the attacker can later take control of your machine at any time while the backdoor is visible to them. So, even standard, simple browsing to infected, unsafe sites can do a lot of harm – invisible harm.

Most of us purchase our own computing devices, such as laptops, tablets, and phones, or get them issued from our office. The device's CPU, memory, and storage are expected to be used by you for personal or official work. If your device's CPU, memory, or storage is used by someone else without your knowledge, it's called **crypto-jacking**. Crypto-jacking by definition is a crime. It's similar to someone using your car without your permission, except that when someone uses my computer CPU, it's not visible and is difficult to track, so it does not generate the same response as if someone just drove my car away for their work. Most coin-mining attackers use or steal your CPU resources without your permission. It's a type of cybercrime.

Now, let's explore the risk of spyware. It derives its name from a *spy* or a secret agent. You must've heard about Hollywood spy blockbusters such as James Bond and Kingsman that show how some countries such as the United States and the United Kingdom recruit spies across the world. A common definition of a spy, as per the Collins dictionary, is "*a person employed by a government or other organization to secretly obtain information on an enemy or competitor.*" So, governments want to keep society safe and hence have teams, units, and divisions created to manage, monitor, and maintain law and order. Let's take a look at some examples (the photos have been taken from `https://www.theguardian.com/culture/gallery/2013/apr/13/10-best-real-life-spies`). This gets more interesting when we bring digital spies into the picture:

Figure 3.5 – Pictorial representation of humans (reel and real) and software spies

So, how is software spyware different than a real-world spy?

If you look up "spy world" on the internet, you will be presented with images of popular spies from the real and reel (film) world around us like the ones shown in the preceding figure. Real spies work differently than reel spies do. If I were to ask you to choose the odd one out in *Figure 3.5*, most probably, you would choose the last one (the phone screen). A spy from the real world evokes certain memories and modus operandi in almost all of us. A software-powered device in our hands could be the home to many such spies, except for their physical presence. The digital spies hidden in our devices can be both friendly and unfriendly, but you need to impart the association you have with real spies to these hidden spies. Let me explain how. Well, a real-world spy would be keeping tabs on the movement of their enemy or competitor constantly. In today's world, all of us are surrounded by various digital gadgets, from mobile phones to cameras in our homes, Amazon Alexa, or other IoT devices in our surroundings, smart TVs, electric cars, and other humans with digital assets such as phones, tablets, and connected IoT devices. In the digital world, if I must keep a tab on my enemy, I will try to get access to all the digital assets they own.

What this means is I can see what my enemy is browsing, what application they use, what time they operate, who they are chatting with, what time they wake up, and when they are asleep, and surrounding cameras can also provide me with a live feed of what my enemy carries, who they are accompanied by, and more information for me to make my next move. Unlike a human spy who will need to have rest and sleep to prevent fatigue, my software spy is awake, full of energy, and working at the same efficacy around the clock. More importantly, my software spy can process information at speed multiple times faster than humans using artificial intelligence. Welcome to the new age of software spies.

It's also interesting how this spyware market has evolved into a business that allows software companies to sell spyware. Spyware is procured by parents to keep an eye on their kids, ensuring they don't fall prey to digital fraud or digital bullying. It's also used by a few employers to keep an eye on their employees. While most countries have privacy laws that govern the legitimacy of spying on children, employees, or other people, it's a practice that is undertaken mostly in the name of national security.

While we are on this topic, it is worth mentioning Pegasus (`https://en.wikipedia.org/wiki/Pegasus_(spyware)`). It's one of the most advanced pieces of spyware in the world and was created by a company in Israel, and it is claimed to be used by many governments and powerful companies on its adversaries. It was primarily aimed to be used by governments to combat terrorism and crime. Over time, it began to be used in other areas, including keeping an eye on political opponents by the parties in power, which resulted in human rights violations and litigations around the world. This resulted in the closure of the company and the making of the Pegasus spyware itself. Well, it might be a closed chapter for this company, but it has not died down in terms of concept. I believe that many new Pegasus tools arrived on the market and they might even be operational today.

The risk associated with applications, or apps, on phones is another that makes it to the top of the list. Let me ask you, how many apps do you have on your phone? More than 10-15? You might even have around 20-50 on your phone. Let's go further – which apps on your phone have access to your microphone and camera? Huh, a tough question that most of us don't know the answer to. I remember installing what is now my favorite news-reading app where, upon installation, the app wanted me to authorize access to my camera and microphone. What the heck? Why should a news-reading app ask for access to my microphone and camera? Well, if I don't give access, can I even install the application on my phone?

One attribute of invisible risks is that it's extremely difficult and not feasible for humans to detect them. The only effective way to detect them is by using advanced software. This software is way smarter than traditional antivirus software. This segment of software detects the behavior of the machine and then, by using pre-identified patterns and artificial intelligence, can detect invisible risks. Currently, they operate under the name of **Endpoint Detection and Response (EDR)**. EDR works in partnership with traditional antivirus and can detect, prevent, and block a whole lot of invisible risks.

This brings me to think about how EDR works. How can EDR detect the intent of the attacker? What playbook does it use, if any, to detect such attacks? If there is a playbook, what if attackers are exploiting the playbook, assuming it's available for public view? What if the attacker uses a playbook to come up with a new technique that is not captured in the playbook? Let me answer these questions one by one.

Yes, there is a playbook that is used by most cybersecurity professionals across the IT industry at large to assess various attack methods used by attackers and create defense techniques. One of the most referenced playbooks is the MITRE ATT&CK framework. It lists, as shown in the following screenshot, what possible methods attackers can use to gain access to mobile phones. Will it be easy to gain access via the browser? Will it be easy to gain illegitimate access to the phone by hiding in a mobile app?

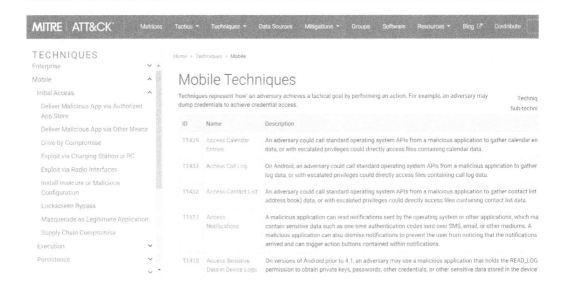

Figure 3.6 – Attack techniques as per standards bodies – MITRE

Assuming that the attacker has access to your mobile OS, what can they access from messages, application data, photos, or other personal data people have on their phones? This brings us to think about what activities the attacker can execute on the phone without requiring a notification asking for approval from the phone owner. Can they delete files, and will that prompt a notification such as *Are you sure you want to delete files?* or will the files just get deleted? All these **tactics, techniques, and procedures** (**TTPs**) are explained in MITRE playbooks that can be used by an organization to develop protection and alerting methods.

At this point, we are now aware of both visible and invisible risks. That brings us to the question, *when does a visible risk become invisible or vice versa?* Is there a way to be proactive against high-impact invisible risks? We'll try and answer this question in the next section.

When does risk become visible?

Digital risks are a priority for consumers and organizations at large. One of the most common ways of thinking is *what is truly a risk?* When can risk be ignored? What mitigation tasks should I plan for risks that have a high probability? What risk can I just ignore and accept since I don't have the mitigation strategy or it's too costly to tackle it, and am I going to make the business decision to accept this risk? Also, at times, we get confused between risk and uncertainty. An easy way to distinguish between the two is to think of them this way: risk is the uncertainty that matters. It matters because it can impact the expected output. So, there will be a lot of uncertainty that will not matter as much and will not fall under the definition of risk.

In the world of digital risks, it's important to understand risk, uncertainties in technology, software, hardware of vulnerability, patched system, bugs in software, digital attacks such as **Denial-of-Service (DoS)** attacks, and the consequences that the digital systems will have in our real physical and digital lives.

I shared a few examples in the previous section while talking about visible risks. These risks come from threats such as phishing attacks, social engineering attacks, and malicious phone applications. The actors creating the threat can be insiders, criminals, and nation states, and can have different motivations, from political agendas to financial gains.

Next, let's look at vulnerabilities. In the digital world, *vulnerable* refers to a weakness, an error, or a bug that can be exploited by attackers. Once exploited, the attacker can either gain authorized access, exfiltrate data, use your computing resources, or change the data to misrepresent things, leading to wrong decisions.

When a risk becomes visible and starts impacting you, I call it a consequence. So, in simple terms, *consequence* is the actual damage that occurs because of digital risk. Depending on the nature and severity of the digital attack, consequences may impact a human or an organization's finances, reputation, compliance status, and business-as-usual operations.

Let's look at a few scenarios where the risks become visible and lead to consequences.

Hit by password phishing

Let's say your password is phished. Your social media password, and possibly your corporate email address, is with the attacker. What are the risks now? What will be the consequence if someone has phished your internet banking username and password? As you can imagine, by using a phishing technique, someone else can operate your digital assets, from social networks to your emails to your bank account. At times in our corporate life, we do share the passwords of certain accounts; these accounts are mostly service accounts with passwords shared between administrators or peers in the same group.

Imagine an attacker with your social media password who is posting on your behalf. Imagine if this post targets a politician, a country, or a law that the attacker doesn't like. Well, they've expressed an opinion in public on your behalf. This kind of attack (where the attacker uses your identity to express views in public that can result in outrage, legal action, or division of society) is a different kind of risk. If you are in a country where the government is proactive in reading, triaging, and acting on such digital posts, then you may be in trouble with a law enforcement officer. Hey, but you did not post it.

While that is true, your credentials were stolen by an attacker. The attacker posted, but how will you explain that to the law enforcement officer? Even if you had to explain it, how will you prove to the officer that it was someone else and not you who posted this message? How will you even collect the evidence? Will this evidence even be accepted by the law enforcement officer? If you take this position, will the law enforcement officer also join you in the investigation or will they assume that you are taking this position to protect yourself and that you were the original person who posted this message?

Let's look at a very simple email usage where an attacker has stolen your credentials via phishing. You get a call from your HR manager saying that you need to come and meet them as one of the female employees of the company received an email containing derogatory messages. Derogatory messages? But you don't even know this person! Possibly, an attacker used your email address after they stole your credentials to send derogatory messages to this individual who has now complained to HR.

Well, this is no different than the earlier scenario where an attacker used your social media profile to post sensitive messages, creating a legal risk for you. In this scenario, you are possibly at risk of losing your employment based on your HR policies. In response, you may have to provide evidence that someone else may have used your credentials or machine or hacked into your account to send this email on your behalf.

As you can see, the credential stolen risk has now been converted into a visible risk, and the consequences of the risk are that you may get dismissed from your employment. In the earlier case with a law enforcement officer, the consequence could be that you end up in jail.

In both these scenarios, there was a digital activity that took place from your account. Your account was used by an attacker after they stole your credentials. They stole your credentials using phishing attacks. Hence, the risk of stolen credentials via phishing attacks became a visible risk and had consequences regarding your reputation, peace of mind, and ability to live a normal, peaceful life.

Now, let's see how a stolen credential could also have consequences of a financial impact. Let's say somebody stole the credentials that you use to log in to your banking site. It may be a banking site, it may be a stock trading account, or it could be a mobile wallet through which you make micropayments.

Use of your credit cards

Credit card fraud happens when somebody makes unauthorized use of a stolen credit card. Globally, millions of credit card numbers are stolen each year, accounting for billions of dollars of unauthorized illegitimate purchases. Credit cards numbers and other critical details are stolen from websites, e-commerce portals that share these credit card numbers for later use, and web browsers where attackers install the keyboard logger, which they use for stealing the credit card details as you type them to make your purchases on various e-commerce websites.

Today, credit cards are more secure than ever, as we have bank regulators, and card providers take considerable efforts in securing transactions. The payment card industry today is governed by a regulation called **Payment Card Industry Data Security Standards** (**PCI DSS**) but still, credit card details getting stolen is common.

A common practice among attackers is to steal credit card numbers, expiry dates, and CVV numbers from any e-commerce website. Today, we have thousands of e-commerce websites selling various products and services. Most of these e-commerce websites use different software databases and cyber security practices to govern the credit card data they have as part of a transaction that you made. What if this e-commerce company's technology, databases, and processes are not strong, and the attacker was able to steal the card details? While you went to the e-commerce portal to buy the product or service, you were tempted by a big discount, but now this invisible risk has become visible and can have consequences.

The consequences are simple: your credit card would be misused by the attacker to buy products and services that they can later redeem into cash or sell to others. The consequence now is who bears this payment that was made without your authorization? Different countries have different regulations.

In the US, federal law limits the liability of a cardholder to $50 in the event of a credit card getting stolen, regardless of the amount charged to the card. Most cardholders will only see this anomaly once they get a statement, and they go through the statement to see whether it is correctly reflecting the authorized transaction they made using the card. In the US, the cardholder has to report such fraudulent transactions within 60 days of receiving the statement. In the US, if the credit card is not physically stolen but the credit card account number and other critical details are stolen and misused, then the cardholders have zero liability for the credit card issues.

In the United Kingdom, the rules are similar where if the physical card was stolen, then you are liable for misuse. Now, why should I be liable for misuse if the card was stolen? Some banks have come out with a novel idea wherein they will make you liable not for the entire amount spent on the card if it's lost, but will cap your liability at a maximum of £50. Some banks have taken this to a level wherein they offer a credit card with zero liability to the original owner in case of misuse of the card by cybersecurity criminals. Some card companies, such as VISA, also use the zero liability features as part of their marketing campaign in selling their credit cards:

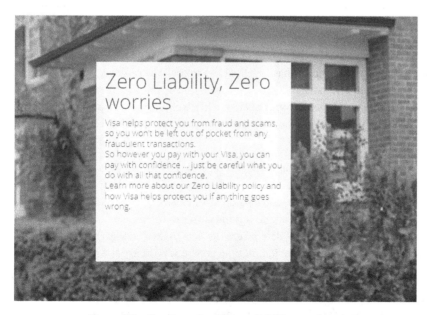

Figure 3.7 – Credit cards with zero liability on misuse

At times, we send photos of a credit card (front and back) to our friends and family members so that they can use it as an exception or in an emergency. What if, while sending the images, the attacker was able to get this information? Now, the seemingly invisible risk becomes visible and has consequences. In the case of a credit card, while the consequences can be financially less or even zero, imagine the time, effort, and mental stress you have as you go through explaining this activity to the credit card company.

The situation becomes worse if it's a debit card. In the case of a credit card, it's impacting your credit limit. In the case of a debit card, it's the actual amount that has gone out as part of an illegitimate transaction. This means that in the case of a debit card, you actually have to not only request the bank to label this transaction illegal but also ask the bank to reimburse you for that money. This is way more difficult than just having a credit limit refused and the credit card company fighting for that amount with the merchant. At least in the case of a credit card, your actual money is safe, so does that mean I should never use a debit card on the internet and secure debit card numbers more strongly than credit cards? The short and long answer is yes.

Ransomware

Like other attacks, ransomware attacks are also on the rise. Individuals and organizations of all sizes are at risk. In organizations, it is mostly finance departments that are at high risk.

In May 2021, a ransomware attack led to the shutdown of America's largest fuel pipeline, named the **Colonial pipeline** hack. In June 2021, another ransomware attack also crippled the world's largest food exporter.

So, what is a ransomware attack? Ransomware attacks encrypt the data on your machine or mobile phone. Ransomware attacks can also encrypt the data that is in your cloud drive, regardless of whether it is being hosted on Amazon, Google, Microsoft, or any other cloud service provider. For a celebrity, a ransomware attack may encrypt the photos and videos that are taken on their phone and ask them for a ransom; in other words, money has to be paid to unlock the same. A very advanced form of a ransomware attack could also have malicious software on the phone that was listening to a private conversation made by a businessman, politician, or a key well-known personality, and it threatens to make that information public unless money is paid to the attacker. The sad reality is that you have to pay to get back control of your data.

Let's look at a small accounting firm or law firm. Most of these businesses involve sensitive to very sensitive data that is shared by their clients. What if a ransomware attack happened and the attacker encrypts all the data and asks for money to be paid for the data to be decrypted back? To make matters worse, what if this entire ransomware attack is done closer to the time of an account filing, or in the case of a law firm, closer to the hearing date where the data plays a critical role? Beware we'll have no time to go through the backups, restore it, and then go about business as usual.

Most organizations rely on the backup of data that they keep in storage vaults in response to ransomware attacks. Organizations also do mock tests regularly to ensure the backup can be used successfully in case of any ransomware attack or human error that leads to loss of data. Whether paying the ransom is the right thing to do is a different question. So, in this case, just the simple act of storing all the accounting data of your client on a cloud drive or saving all the images that are private on your phone, or just having all the case notes for a law firm in a document becomes risky if the access to such storage is gained by ransomware or malicious software.

Invading privacy and extortion/phone spying

Politicians, movie stars, and well-known celebrities are at high risk. When it comes to taking digital revenge, any one of us (whether we have just started school or college or have just joined the workplace) is at high risk. Imagine privacy concerns in the context of cyber security and digital risks. Shortly, I will describe how a ransomware attack could happen in a company where we have a finance department or a legal department working on a case on behalf of a client. I will also share the risks individuals have if they are politicians or celebrities regarding their photos and voice.

Let's assume you're a business leader. You are discussing your business strategy with your team. Your phone is listening to the entire conversation while it silently rests in your pocket. Your phone not only listens but even has the plan to send a copy of the conversation to an attacker's email who had maliciously gained access to your phone and planted this malware. The question is, is it possible for a mobile phone kept in your pocket to listen and pass this sensitive conversation to attackers? The short answer is yes.

Imagine you are a celebrity and you are taking pictures on your phone or exchanging messages on WhatsApp or other platforms with another potential celebrity. What if your phone was compromised and it was sharing all this information with an attacker who would then sell this information and private dialog to magazines that would cover that? In this case, your information is stolen without your consent and your privacy is invaded for an attacker to demand a ransom for you. Or the attacker could have a different business model where they are selling this information to a publisher. Nowadays, on the dark web, people are also interested in paying so that they can get this information and possibly pictures – mostly private, explicit photographs of certain celebrities – and the payments are made using a type of digital currency called cryptocurrency.

In the previous section, we learned about how small organizations and individuals are impacted by digital risks. The digital risks that were invisible suddenly become visible and have consequences. These consequences are both financial and non-financial. Now, let's look at some scenarios where we are employees working in a large organization, and how seemingly normal activities could put our organization at risk and also put us at risk.

Let's say I'm working in a consulting organization. We work with lots of **intellectual property** (**IP**) across the documents, software, and processes of the organization. Most large organizations have multiple vendors and suppliers who work and make copies of these IPs to perform their work responsibilities. As part of the normal routine work I do as a consultant, I send a lot of information in documents, PDFs, emails, and video recordings of certain meetings that we have during the current pandemic times to our suppliers and partners.

As part of sending documents to the suppliers, I had to send a couple of files that contained sensitive information about our upcoming product. It contained confidential information on the product design and also the list of features that our product will have. In the comments of one such document, there were also references to competition and how a product could be positioned well against the competition documentation and public notes. What if this document got into the hands of the competition?

What if this document got to a partner and now, based on the features that we are not creating, the partner is looking to create their own IP? As you can see, I, as a consultant, was doing my normal job of sharing information with my supplier or partner so that both of us could work together on the common goal of creating this product. What I did was unknowingly pass such sensitive information to the partner and also capabilities that I as a consultant will not be developing. This gave the partner critical information if they wish to exploit and build this IP.

This was a human error of giving this critical information to my partner or supplier. Imagine if this information was not given by me but was stolen by malware and then given to the attacker. This is a data exfiltration attack that now puts the organization at a visible digital risk. Data exfiltration malware refers to malicious software that is designed to steal sensitive information from a targeted computer or network and then transmit it to an external source, such as a remote server controlled by the attacker. As you saw in the preceding scenarios, the normal use of technology can be risky. It's also very difficult to assess when a hidden risk can become a real risk. It's also difficult to measure the exact impact that a real risk poses to an individual or a small company. The consequence of the risk will impact your reputation, finances, and even mental well-being.

Summary

The digital transformation of humans as consumers, employees, or employers is well underway and is impacting both personal and professional lives. While the rate of digital adoption will be different for each of us, I believe it brings different kinds of risks. Some of these risks are visible, while others become visible over a certain period, leaving an impact with varied consequences. In this chapter, we reviewed various scenarios and the need to stay alert. Most of us are still unprepared to deal with different types of attacks and are, at best, aiming to mitigate the risk rather than preventing attacks in the first instance.

In the next chapter, we'll review the risk that organizations, multinational corporations, and governments face in current times. The pandemic brought in a new way of working – working from home. We'll explore how an organization can balance productivity and risks when employees choose to work remotely from their home or a café.

4

Remote Working and the Element of Trust

After many staggered starts, the full-blown era of remote working is now here. Remote working is a working style that allows professionals to work outside of a traditional office environment.

Is working in an office different from working remotely? Can everyone work remotely? Is it more productive to work remotely? Well, remote working is based on the concept that work does not need to be done in a specific place.

Can I work from home on some days and work from the office on other days? Is that considered to be remote working? If **working from home (WFH)** means remote working, can I work from my friend's home, my parents' home, or my vacation home? As an employer, do you really care about where I work as long as I'm working from a remote place and being productive? People have the flexibility to design their days so that their professional and personal lives can be experienced to their fullest potential and coexist peacefully.

In recent years (especially since COVID-19), WFH has become a term known to almost every person on the planet. Even my parents today know what WFH means. Most organizations are now extending it from WFH to **working from anywhere (WFA)**.

In this chapter, let's explore what WFH means. We'll learn about the professions in which you can feasibly work from home. We'll look at the element of trust, which is needed if you are WFH, and the kinds of risks, both visible and non-visible, that are present when you are WFH.

We'll cover the following topics in this chapter:

- Remote working – not new for everyone
- The pandemic and remote working
- View of remote working for various industries
- Risks to organizations

Remote working – not new for everyone

Remote working is not a new idea. People have been working remotely for centuries in various forms, such as telecommuting, freelancing, and WFH. The concept of remote work has been around for a long time, but the technological advancements in recent years have made it more prevalent and accessible for a wider range of industries and job types.

I remember that I used to tell my parents, "*I have a meeting in the afternoon so I will go to the office to attend it.*" During my office hours, I would stay online, check my email, and work on the documents I needed to work on as if I were physically sitting in the office. I used to tell myself that I had a meeting around noon, so I would go to the office later and finish my work at home, which would allow me to beat the rush-hour traffic and go directly into the meeting. There is evidence to suggest that remote working can increase productivity. Studies have shown that remote workers tend to work longer hours and take fewer breaks than office-based workers. Additionally, the lack of a commute can give remote workers more time to focus on their work. In my experience, productivity will depend on factors such as the individual's work habits, the company's culture, and the tools and resources available to remote workers.

A lot of my peers have also been working remotely since 2000. I can work remotely due to the nature of my work. I remember a lot of my colleagues who worked in sales spent almost 4 out of 5 days out in the field. They never had to come to the office at 8:00 A.M. and then go home at 6:00 P.M. Most of these sales team members used to come to the office only when there was a meeting called by the sales leader. Why should a salesperson even be seen in the office? A salesperson should be with the customers, selling the products. The same applies to knowledge workers and consultants (such as me back then). It can certainly be said that certain roles in some companies have always been able to work partially from home, more so in the last few years with the pandemic. Obviously, there are certain roles, such as taxi driver or airline pilot, in which you can't choose to work from home. The same can also be said about roles such as that of a surgeon or a factory blue-collar worker. Advances in AI in medicine and automation in assembly lines might already be changing this in some of these fields, but we will leave them out of scope here.

Even before a lot of us started working remotely, a NASA engineer by the name of Jack Nilles laid the foundation for modern remote working when he coined the term **telecommuting** in 1973.

Now, what if my work was manufacturing a car? Can I manufacture a car from my home? Certainly not today, but maybe one day you might be able to! The point here is that not everybody can work from home. Can a doctor work from home? Can a medical surgeon perform surgery from home? Not as of yet.

So, a small minority of us were WFH in the last few years. In my company at least, technologies such as Skype, Lync, which was quickly renamed Skype for Business, and now Microsoft Teams have advanced so well across the organization that I am beginning to question whether we really need to be together in an office to collaborate and do our work. I think we got a loud and clear answer during the pandemic lockdowns.

WFH can be productive if you have a good working environment created at your home. As I think back, I can remember various office-like setups that I created in different cities in the last 20 years.

Pandemic and remote working

A **virtual private network** (**VPN**) is a technology that allows users to securely access a private network and share data remotely through public networks. VPNs are often used by remote workers to securely connect to their company's network while WFH. This allows them to access files, applications, and other resources as if they were physically on the company's network. Additionally, a VPN can also help to protect the user's internet connection from potential threats, such as hackers. Many organizations had a similar setup and they came under stress during the pandemic. Most organizations had created this VPN or WFA environment for their executives or on-field employees. The capacity of this infrastructure was not meant to serve everyone working from a remote location. Boom! Along came the pandemic, and the limited capacity of this remote working infrastructure was put to a stringent test. Most companies struggled to issue laptops to all their employees and make them ready for remote work.

In a survey of 100 North American C-suite executives across different company sizes, as many as 95% of all CIO respondents reported that their IT organizations have been bogged down by the inability to efficiently provide a remote working infrastructure since the COVID-19 pandemic accelerated a shift to a remote workforce. The story was no different (rather, much worse) in other parts of the globe (`https://www.computerweekly.com/news/252485267/95-of-North-American-CIOs-report-remote-work-issues-during-lockdown`).

The COVID-19 crisis has opened senior leaders' minds to the idea of adopting WFA for all, or at least a part, of their workforce. A quick Google search will show that most large organizations (including Microsoft, Facebook, Google, TCS, and Accenture) have come up with liberal and sometimes mandatory remote working policies. Even customer-facing industries, such as banks, were forced to assess how they could maintain business as usual while employees worked remotely.

There was one thing that was essentially force-fed to organizations. Executives went to formal board meetings digitally, and critical reviews of businesses were done while executives worked from remote locations. Key decisions were taken without sitting face to face. No corridor talk, no break-time conversations, no side catch-ups – meetings for running and growing the company were all held digitally. It was unexpected, unthinkable, and impracticable what COVID-19 forced upon executives around the world. I can't resist sharing this viral image that circulated very shortly after the pandemic started and will quite possibly be remembered for a long time:

Figure 4.1 – A cartoon targeting the digital transformation of companies after COVID

While we can spend all day and night debating the advantages and disadvantages of remote working, the intent of this conversation is not to make a case for or against remote working that further accelerates the adoption of digitization. My interest lies in sharing the risks, both visible and invisible, with you that can have consequences for both employers and employees as a result of working remotely. From an unsecured Wi-Fi connection to an unpatched machine, or even a friend's machine being used in an emergency for a short time to check email, and many more such scenarios, these can put both organizations and employees at risk.

In the preceding sections, I have tried to paint a picture of how remote working has been around for quite some time, albeit anecdotally. In the next section, we shall look at how the ability to work remotely is also dependent on the industry and the kind of work that is expected from the employee.

View of remote working for various industries

The viability of remote working varies across industries. Some industries, such as technology and consulting, have been able to successfully implement remote work for the majority of their employees. Other industries, such as retail and hospitality, may have more difficulty transitioning to remote work due to the nature of their business operations. Additionally, some jobs within a given industry may be more suitable for remote work than others. Overall, the ability to successfully implement remote work will depend on a variety of factors, including the specific industry and job function, as well as the technology and processes in place to support remote work. It is a no-brainer that it is easy for a knowledge worker (who primarily works on a computer) to work from home versus a manufacturing worker who must go to the factory or a taxi driver who must drive the taxi on the road and take passengers from one location to another.

When talking about the manufacturing industry, the physical nature of manufacturing jobs makes switching to remote work a challenge. A survey done by Pew Research showed the employee viewpoint on effective remote working. The following graph can be found at https://www.weforum.org/:

Ability to telework varies widely across industries

% of employed adults who say for the most part, the responsibilities of their job can be done from home, by industry

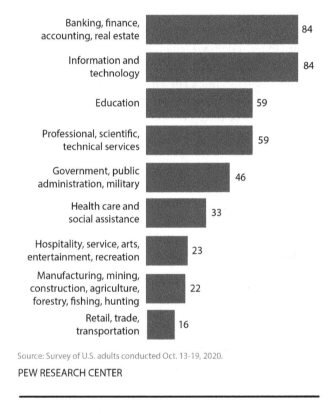

Source: Survey of U.S. adults conducted Oct. 13-19, 2020.
PEW RESEARCH CENTER

Figure 4.2 – Industry-level view of the ability to do work remotely

Let me share my views, sector by sector, on this report. As I stated earlier, it is less feasible for a taxi driver to work from home. As you can see in *Figure 4.2*, the retail, trade, and transportation sectors scored the lowest in the ability to work remotely. If I let my imagination run wild into the future, could we not have a world where a taxi is operated by a human sitting remotely? Do I really need a taxi driver to be sitting inside the taxi? Ignore the driverless concept for a while and this seems to be feasible as long as there is good connectivity between the car and the remote taxi operator. If I switch to manufacturing, from the mining sector to the automobile industry and from oil rigs to fishing firms, we have automated systems and machines that work, and humans are just behind the machines, operating them with buttons and levers. Could we not have cameras and remote operations for the same set of machines to enable workers to operate them remotely, as long as the network and bandwidth were good? This might not be normal, but neither was a mobile phone just 3 decades ago.

As you move up the sectors and industries, from government administration to hospitality and education, more and more digitization opens up the chances of delivering, manufacturing, or servicing/maintaining the products and services remotely. We will cover the future more in the last chapter.

Most parents, including me, never really believed in remote teaching via video lectures. Well, guess what? Students across the globe from K1-K12, undergraduates, and graduates to anyone studying for a professional certification had to learn skills via remote teaching. It had its own learning curve. I was amazed to witness my 7-year-old daughter attending her ballet class perfectly well, along with a few other children, all remotely on Zoom meetings. For a few months, it was hard to believe the change and now it's opening up opportunities globally across many sectors. Where there is smoke, there is fire. Where there is digitization, there will be risks. Let's look at the risks that are faced by organizations from insiders across industries in the next section.

Risks to organizations

Today, as we speak to risk professionals across corporations, they talk about cybersecurity risks to an organization. I call them *outside-in risks*, that is, risks from outsiders, attackers, and hackers, but what about *inside risks*, or risks from insiders? This is the risk this new remote working environment creates as employees work from remote places.

What if, while working from a remote location, my finance executive passes sensitive information to my competitors for personal gain? What if sensitive information about company performance is used by a family member of my CFO with or without their awareness? How confident are you that the subcontractor that you are paying as a retainer is only working for you and not double-selling their time working from a remote office? How comfortable and confident are you that critical information with your partner working from a remote location will not be shared, leaked, or forwarded to your competition?

When employees were working from an office, all sensitive information resided in office networks, with office cameras and security staff providing both digital and face-to-face surveillance.

Even before the pandemic, disgruntled employees driven by personal agendas worked to infiltrate company data. Remote working made this mission easier. Cybercriminals got soft targets to prey on home networks, which often lack strong cybersecurity defense strategies. For a company's IT team, it's often impossible to configure, access, and monitor the home networks of employees. Remote working is not about just the home, it's also about working from anywhere. This includes a café, hotel, airport, friend's place, or long-stay apartment, where a network is shared across various unknown residents.

Not only have we gone remote but we have also started to adopt technologies such as Teams, link-based sharing on SharePoint, WhatsApp-based file transfers, Zoom, Google Drive sharing, and phone-to-phone video calls that may not be under the direct control of your corporate IT and security teams. Bringing your own device and working on your personal computer or mobile phone makes it even more difficult to assess the risks your devices pose to an organization's network and computing resources. The term popularly used for these technologies was *modern workplace*, which seems to now have evolved to *modern remote workplace*.

A quick Google search on remote worker surveillance software will come up with scores of software working under the cover of employee productivity software. It's no different than parents using spy apps to monitor their kids' activities. Most countries have legislation and regulations, as this invades privacy. Do I really want my employer to have 24/7 access to my camera and what I am doing on my personal computing device?

The answer to that question is slightly complex. We do not want to give up our privacy, and no employee would welcome a culture of surveillance. That does not take away from the need for employers to be cognizant of the outsized risks that a few employees can advertently or inadvertently expose the organization to. I would like to give an example of a tool that is trying to balance these two ends of the equation – risk management for the company while ensuring privacy and non-surveillance for employees – the Microsoft Insider Risk Management solution. This unified solution can paint the risk profile of any of the employees based on their actions around sensitive information across Office 365 emails, cloud storage locations, device endpoints, and portals across mobile, desktops, or laptops. These activities are scored anonymously, regardless of whether an employee is working from a remote or office location. The solution also provides the appropriate analysts as empowered by the organization with investigation and triage workflows for identifying potential risks and taking action to mitigate those risks. If you are a prospective employee, this can be a natural source of anxiety: "*Is my employer going to monitor all my actions across sites and devices?*" In the answer to this lies balance – the tool, by default, does not store any of these activities for any employee and is completely opt-in (the organization will have to choose to activate it and specify appropriate investigators for using this tool). The tool operates on the principles of triggers and indicators. Trigger events are a well-defined set of activities that bring a user into the scope of the evaluation. Indicators are activities that indicate risk.

Let's say an employee sent a sensitive file to someone outside the company. Based on configurable settings, this might qualify for the tool to start evaluating the employee's overall actions in the last few days and for the next few days (the trigger). During the evaluation, the name of the employee can be anonymized, and only after there is a sufficient amount of evidence of wrongdoing (via indicators) can the security admin de-anonymize the name and proceed with the organizationally agreed next steps.

What I like the most about this software is the Analytics scan, which can be used to identify potential insider risks. It detects user activities across various locations and provides anonymized results and recommends what policies to set up to ensure you can mitigate risks. Generally, when an organization begins its internal risk analysis journey, they start with a lack of knowledge. They have concerns, but no specific patterns that they can start to address.

Additionally, the internal stakeholders, such as the privacy team and the human resources division, also want to understand the need for a tool that deals with their own teammates and how it works.

Insider Risk Manager in Microsoft 365 paints anonymous pictures for all these stakeholders. Without naming any employee, it points to patterns of concern in the organization as a whole.

The solution also works on preconfigured playbooks, also known as templates. Some templates that the software provides include data theft on the part of departing or disgruntled employees, general data leaks, and security policy violations. It also gives you the flexibility to configure indicators for suspected users to track potential risks. As we mentioned previously, indicators are a list of activities that are deemed risky. As an organization, this can vary from source to source. Hence, the tool provides you with a host of actions to configure from. For organizations at an early stage of evolution in interpreting their insider risk landscape, the tool also provides default settings and threshold-based AI and ML.

It's amazing to see how AI can predict across millions of indicators coming from remote workers. The following screenshot shows how this software can correlate the data exfiltration from a remote worker and paint a timeline of activities that the employee undertook to cause risk:

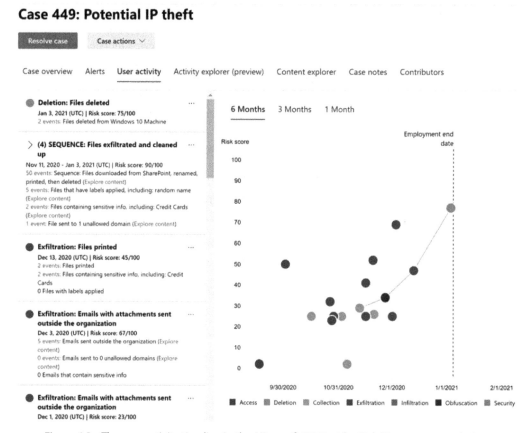

Figure 4.3 – The user activity timeline in the Microsoft 365 Insider Risk Management solution

On the left-hand side of the figure, you can see the list of activities that are considered risky. These are based on the indicators that the organization's security admin chose at policy setup.

That ends a brief foray into a solution that helps you manage risks that remote work exposes your organization to. We chose one of the Microsoft solutions for illustration because of familiarity and access, but we'd be remiss if we did not tell you that there are solutions of various competitive capabilities in the same area, and a quick web search on `Insider Risk Management software` would lead you to any of them.

Summary

In this chapter, we looked at the evolution of the remote workplace and the risks associated with it. During the COVID-19 pandemic, many companies implemented remote work policies to ensure the safety of their employees. This has allowed employees to continue WFH, rather than going into an office. The pandemic has also accelerated the trend of remote work, as many companies have realized the benefits it can provide, such as increased productivity and cost savings. After the pandemic, remote work will likely continue to be a popular option for many companies and employees. Some companies may continue to have a fully remote workforce, while others may adopt a hybrid model where employees have the option to work remotely or in the office. With anecdotal examples, I depicted how remote work has always been around in some shape or form, but the scale and scope of remote work have just multiplied since the pandemic. We also got a view of the industry-specific trends and capabilities around remote work. We ended the chapter with a view into the risks that remote work exposes companies to and looked at one of Microsoft's solutions to mitigate them.

In the next chapter, we will get a quick introduction to another popular concept around cybersecurity – zero trust.

5

The Emergence of Zero Trust and Risk Equation

Every organization these days is a technology organization. Due to our increased usage of technology, a bank is no longer considered a traditional bank but is considered a tech company masquerading as a bank. These days, our digital business has no defined perimeter as most employees prefer to work from home.

Organizations have increasingly started adopting the *Zero Trust* architecture framework; in this chapter, we will discuss what Zero Trust architecture is all about and how Zero Trust can help manage risk and compliance.

In this chapter, we are going to cover the following topics:

- An example of the Zero Trust concept in a real-life situation
- The role of Zero Trust in digital transformation
- Modern principles of Zero Trust
- The intersection of Zero Trust with digital risk and compliance

Zero Trust in real life

Life as a parent is never easy, and it's always a roller coaster ride with a share of ups and downs. After a hectic week of working remotely, managing virtual events, and minding two teenagers, you're looking forward to a quiet weekend. As the weekend approaches, you realize it is your elder daughter's 16th birthday, and you couldn't be happier. Suddenly your smile fades as you start to imagine a bunch of teenagers being let loose in your house after weeks of a recent lockdown. Being the excellent parent you are, you take deep breaths, process your feelings, and discuss the rules of engagement regarding your daughter's sweet 16 birthday bash.

The big day arrives. Your daughter is having a ball while you're overwhelmed by the number of invitees. You're watching and inspecting every teen that enters the house like a hawk. Now, being the security geek that you are, you've got the best-in-class wire mesh fence, the latest and the most incredible outdoor home security cameras, and the most robust locks on the front door. However, once you let someone in the house through the front door, they have access to all levels, including your private space, upstairs, along with your memorabilia. You get the picture.

This story describes our traditional approach to security, which is largely perimeter-based. With this approach to security, traffic is divided into trusted and untrusted. Traffic on the outside of your organization's perimeter or firewall is untrusted, and everything on the inside is trusted. This approach places much inherent trust in the traffic within the organization's boundaries, which leaves us at risk in our house scenario if one of the trusted insiders you have welcomed turns malicious.

Our digital businesses today have no defined perimeter. We've adopted the cloud for agility, and our workforce is mainly mobile. We operate through every channel and device that customers use to connect to us. Our partners interact with our data and services remotely. This leads to the perimeter-based approach to security becoming increasingly ineffective. The modern way of working requires a set of security principles in line with the Zero Trust concept and its ability to support the new digital transformation.

Zero Trust is a new digital transformation

While a fundamental transformation in risk management and security has been well underway for a while now, the global pandemic put all of us on an accelerated path and bought forward many of the toughest challenges right to the surface. Our traditional perimeter-based network and security models couldn't adapt and, for many, the existing infrastructure was not ready for such a significant shift to remote work.

Many organizations now need to support permanent hybrid work models in which employees are opting to work from home; at the same time, security leaders are faced with the challenge of securing resources that are beyond the corporate network.

As employees transition to new work models, they're bringing more devices to add to the increase in endpoint diversity that organizations were already seeing.

With all this, the digital estate continues to grow, making visibility and threat detection and response across multiple platforms and clouds challenging. Security teams need more visibility and struggle with threat detection and response across multiple clouds and platforms.

This diversity and siloed solutions make rolling out new security controls difficult and inconsistent.

Bad actors such as malicious users and attackers who want to take advantage of security weaknesses and vulnerabilities know this, and they're exploiting these vulnerabilities.

Finally, there is still work to be done to optimize networks to prevent lateral movement for many organizations.

Lesson learned from a global pandemic

In December 2019, when Wuhan officials reported the first case of viral pneumonia, this news came entirely out of the blue. No one anticipated how our world would be completely turned around; it has been 24 months, and we are still combatting new variants of this virus. Just out of the first variant of the virus, Australia invoked the biosecurity screening of arrivals from Wuhan. This is very similar to how we've traditionally relied on traffic filtering as our first line of defense when defending against cyber threats.

By the end of March 2020, this virus had spread across 170 countries, impacting over 750,000 lives and countless livelihoods. This was when most states in Australia had adopted a series of measures to dampen the spread of the virus, including border control, supervised quarantine for arriving travelers, the expansion of testing services, and contact tracing. Over-reliance on biosecurity screening (or traffic filtering – protection) isn't very effective on its own unless coupled with tracking via COVID Safe (or event monitoring – detection), contact tracing (or incident management – response), and so on. These measures are very similar to how businesses adopt defense-in-depth to strengthen their security capabilities. This is by layering controls by placement, that is, network, application, host, and so on, or by function, that is, detect, prevent, and respond.

In November 2020, we first heard of the term *zero-COVID*, which highlighted a strategy of **Find**, **Test**, **Trace**, **Isolate, and Support** (**FTTIS**). South Korea, Singapore, Vietnam, and New Zealand adopted this strategy with varying interpretations. Our five-pronged approach to zero-COVID in Australia includes social distancing, testing, contact tracing, lockdowns, and vaccination. Similarly, Zero Trust centers around the five foundational pillars: data at the center, users, workloads in the form of physical/virtual servers or applications/systems, networks, and devices. These five foundational pillars of Zero Trust are enveloped with a layer of detect, respond, and recover capabilities.

Most complicated problems, such as the pandemic, never have a silver bullet solution. In parallel to vaccinating the population, enforcing supervised quarantine for inbound travelers who could be the carriers of new variants is also essential.

Zero Trust isn't a single solution that can be bought and deployed along similar lines. Zero Trust is a journey and not an end state as cyber threats keep evolving, similar to our struggles with evolving strains of the virus, such as Delta.

Zero Trust is a proactive approach to security that uses adaptive controls and continuous verification to prevent and respond to threats more quickly and efficiently. It is a timely approach to addressing the cybersecurity risk and challenges originating from the rise in remote working, the proliferation of personal devices, and obsolete physical security perimeters.

Chief information security officers (**CISOs**) and **chief risk officers** (**CROs**), as key trust and risk ambassadors of the business within an organization, should consider that, unlike their traditional belief of *trust but verify first*, Zero Trust dictates that anything out of a defined scope is insecure unless user or application authentication and authorization is completed according to standard company security policy. In other words, it religiously promotes the concept of *never trust, always verify*.

Modern principles of Zero Trust

Forrester developed the Zero Trust concept and framework over a decade to enhance the security posture and architecture to reduce cyber risks and threats by implying trust, explicit authentication, and authorization. To ensure we are on the same page, C-Suite should not consider Zero Trust as a product or an off-the-shelf tool. Instead, it is more of a concept and requires revisiting the organizational architecture, policies and standards, operation model, and most importantly, culture.

The **National Institute of Standards and Technology** (**NIST**) published a guideline in its 800-207 Zero Trust Framework document outline as follows:

1. Defining data and computing services as assets and resources.
2. All communications among different resources and geo-locations must be hardened.
3. Individuals' access to the enterprise should be authenticated and authorized on a session basis with defined session time expiry.
4. User monitoring and access management are beyond traditional log monitoring and behavior and other factors such as application and service state, client identity, and so on play a significant role in resource and access management.
5. Monitoring enterprise assets include integrity and security postures.
6. Granting access takes place only once authentication and authorization are successfully completed.

The NIST approach to Zero Trust architecture

Zero Trust architecture requires planning and connecting different elements such as a **policy engine** (**PE**), **policy administrator** (**PA**), and **policy enforcement point** (**PEP**) and building relationships among different stakeholders, defining workflows, rules, and responsibilities, and defining access policy criteria.

To build an access policy, a number of inputs are required, as described in *Figure 5.1*:

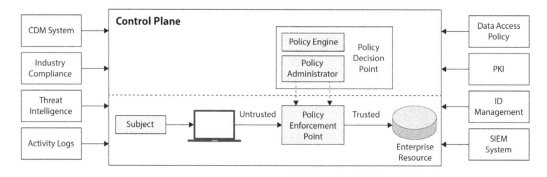

Figure 5.1 – Zero Trust architecture components and inputs [Source: NIST SP 800-207]

To build a successful Zero Trust architecture and reduce the cyber risk level, it is required to obtain an accurate understanding of assets and resources, data flows, and orchestration. Having the mentioned knowledge and other inputs as discussed earlier allows the PE to constantly fine-tune its automated risk assessment.

The Sunburst attack example

The traditional belief has always proved that *trust but verify first* has failed as demonstrated in many cyber incidents. That's the main reason we constantly discuss *never trust and always verify*.

One of the significant cyberattacks that took place in 2021 was the software supply chain attack against SolarWinds, aka the *Sunburst attack*.

In this attack, the hackers used a method known as a supply chain attack to insert malicious code into the Orion system. The compromised service account led to lateral movement and the takeover of the entire network.

As a lesson learned, applying the least privileged philosophy across all organizations, including service and application IDs, and removing unnecessary administrative accounts, is one of the key controls that need to be applied to the Zero Trust architecture environment. Such privileged accounts behaviors need to be documented to assist the PE with identifying the known behavior from malicious and anomaly behaviors and escalating or triggering an alert.

Nowadays, cyber incidents and data breaches are not just limited to regulatory fines or financials and reputational damages, but rather impact the job security of leadership and executive-level staff as well. According to a Kaspersky Lab and B2B International report, 32% of data breaches lead to the firing of senior management and relevant teams in North America. Hence, the governance, risk, and compliance team should be more vigilant to identify the risk and escalate it to the management to help them make an informed decision.

CISOs and security leadership teams need to enhance their information security awareness and training and define more stringent policies against those offender employees that fail in phishing assessments or actual phishing incidents.

Zero Trust across the digital estate

To effectively govern and comply with security requirements, enhance your security posture, and minimize cyber risks, the Zero Trust architecture needs to be deployed across the entire organization and integrated with all security tools and applied security best practices.

To be able to reach the preceding objectives, the following six key elements play a significant role:

- **User authentication and authorization**: Access to resources (data, a network, an application, etc.) should be at the entry point along which each request is verified (authenticated and authorized) at all times. Just because a user is from HR doesn't necessarily guarantee that they need to obtain access to the HR application or data. Also, not all HR staff are authorized to perform a particular action. Hence, user identity and authorization levels need to be assessed and verified.

- **Device verification**: The device's security posture and fingerprints help identify whether the computing device that is trying to access the organization's assets is secure and complies with security requirements. Imagine a malicious user attempting to access certain sensitive data via stolen credentials, the device verification helps to indicate whether it's a known device, whether it has the required security controls (certificate, security agents, and patches), or whether it has the right IP address and geolocation, and review the previous known behavior. Failed device verification could prevent unauthorized or risky access.

- **Data protection**: Data needs to be protected at all three states in transit, in process, and at rest, and the required security controls such as user authentication and authorization, data classification, encryption, and many more need to be applied based on the data criticality level.

- **Application hardening**: Application security attracts more attention these days, and many organizations are investing a larger budget in enhancing their application security architecture, secure design, business logic, and monitoring application behavior to mitigate malicious activity.

- **On-premises and cloud infrastructure**: On-premises and cloud environments need to be secured following security standards, frameworks, and best practices.

- **Network security monitoring and governance**: Capture user, application, and system behavior and monitor their activity using automated and orchestrated tools integrated with threat intelligence. Also, ensure that network architecture is segmented strategically based on the sensitivity and criticality of the assets and departments where required.

The following figure shows the high-level Zero Trust access architecture with the security policy enforcement engine at the center. In the case of Microsoft technology, Microsoft Azure Conditional Access Engine is used as a central security policy enforcement tool to apply access across all the resources:

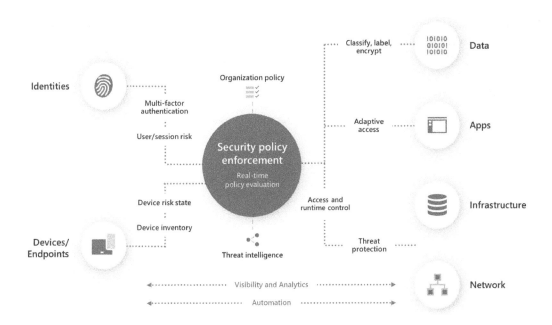

Figure 5.2 – Zero Trust access architecture overview

Microsoft Azure Conditional Access Engine ties all the resources together with the use of the following:

- **Policy-driven access**: Modern micro-segmentation means more than networks, and we also require gate access based on role, location, behavior patterns, data sensitivity, client application, and device security. Ensure all of the policy is automatically enforced at the time of access and continuously throughout the session where possible.

- **Automated threat detection and response**: Telemetry from the preceding systems must be processed and acted on automatically. Attacks happen at cloud speed – your defense systems must act at cloud speed as well, and humans just can't react quickly enough, so integrate intelligence with the policy-based response for real-time protection.

Example of controlling access with intelligent policies and continuous risk assessment

Due to the principle of *never trust, always verify*, only authenticated and authorized users with the right device and security compliance could access the assets locally or remotely. Microsoft Azure Active Directory using **Microsoft Conditional Access** (MCA) enables organizations to define, configure, and fine-tune their access policies using user info, geo-locations, device fingerprints, and risk severity levels based on user access.

Conditional access gives you the power to enforce the core principle of Zero Trust—never trust, always verify. The Zero Trust security model relies on a security policy engine to make access decisions you can enforce throughout the digital estate. MCA fine-tunes the access policy and enforces (deny, access, and redirect to the MFA page) users contextually based on the user, device, location, app permissions, data sensitivity, session risk information, and more.

That gives you better control over how users access corporate resources, as demonstrated in *Figure 5.3*:

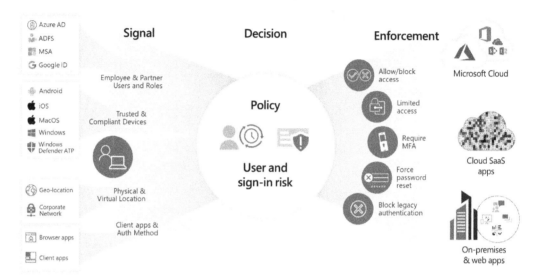

Figure 5.3 – Resource access control with a PE

Zero Trust makes compliance easier

"The objective of effective compliance is to reduce the likelihood of risk, and the same is true of Zero Trust" ~ Abbas Kudrati

The rising number of privacy and compliance standards has increased anxiety among many business owners. Between the APRA CP234 and the Privacy Act, PCI-DSS and HIPAA, the EU's GDPR and Mexico's Protection of Personal Data Law, and **California's Consumer Privacy Law** (**CCPA**) and New York's Personal Privacy Law, these standards have teeth and bite back with stiff penalties for companies that fail to adhere to them.

Zero Trust nowadays plays a heroic role by assisting in reducing cyber risk and compliance requirements and increasing the confidence level to migrate to single or multi-cloud environments.

Now, let's look at how Forrester defined two essential elements, Isolation and Monitoring, and the five strategic process steps to achieve Zero Trust concepts.

Isolation

While Forrester uses the *Isolation* terminology, this is not a new concept in information and IT security words and it has been known for years as *Segmentation*, which is referenced by many compliance standards such as NIST, ISO, and PCI-DSS, to name a few. Isolation or Segmentation dictates that assets need to be categorized based on their sensitivity and criticality and treated differently in different network blocks, storages, or access levels. This approach allows you to define the access level at the block level and once something is compromised, other segments are not easily accessible to intruders.

Defining the criticality level and baseline similar to any asset management standards is required at the initial stage and based on the criticality level, the security controls, tools, and capabilities need to be defined.

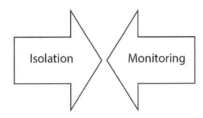

Figure 5.4 – Elements of Zero Trust compliance

Monitoring and visibility

Once the baseline is defined, a gap assessment is required to be conducted to understand the current setup visibility and what security tools and architecture gaps need to be defined. Based on this gap assessment, the roadmap to achieving the required compliance level and successful Zero Trust will be defined. The ongoing and historical visibility allows real-time as well as historical monitoring of user, device, network, and application activity.

As the visibility increases in a more granular manner, enforcing access policies in the next step will be more feasible.

Let's look at how compliance is mapped with Forrester's five steps of Zero Trust networking:

Forrester's Five Steps of Zero Trust Networking	Description
Discover and identify your data and assets	You can't protect what you can't monitor. The first step is to discover organizational assets and assess their criticality level. Identified assets will be the starting point in defining the scope and deploying Zero Trust architecture.
Data flow mapping	One of the key elements of Zero Trust is to identify user behavior and abnormal activity. Knowing the normal data flow allows us to define and indicate the abnormal activity at the user, domain, and process levels.
Creating and defining micro-segmentation	Based on the discovered and identified assets and categorizing their criticality levels and data flows, the right segmentation should be defined with relevant security controls and an access policy.
Continuous monitoring and visibility using an analytics engine	As monitoring is one of the key controls, all behaviors and activities across the network should be logged and monitored at a granular level. These historical flows of logs and alerts become compliance assurance, which proves we are genuinely compliant over time and serves in compliance validation.
Implement and use **Security Automation, Orchestration, and Response (SOAR)**	By leveraging machine learning and artificial intelligence, identifying and responding to the events will be faster with minimum manual operation level and human resource requirements.

Table 5.1 – Forrester's five steps of Zero Trust networking

Zero Trust requires real-time monitoring to obtain granular behavior details, assess each access request, and if they adhere to the defined policy engine, access will be granted. By having this close and stringent monitoring in place, user authentication and authorization, device security compliance, and enforcement based on the risk assessment will be applied in real time. Having successful risk management practices requires governing and auditing the process dynamically and taking action on high-risk items with a higher priority, which positively affects the security posture and risk maturity level of an organization. Not to mention, user awareness and training programs also assist organizations to meet the skill gaps and minimize human errors.

Summary

Out of many use cases of Zero Trust, an organization can achieve better compliance and risk management using this modern architecture and by implementing the principles of Zero Trust.

The Zero Trust framework is the right approach to risk management and the best way to protect your organization in the future as it allows a significant improvement in digital monitoring (governance) and reduction in risk with an increased maturity in an organization's ability to meet other regulatory compliance frameworks.

In the next chapter, we will look at various types of risks and how these risks may impact the overall risk exposure and risk appetite/tolerance of an organization.

Part 2: Risk Redefined at Work

This part focuses on the changing nature of risk in the workplace. It covers topics such as human risk at work, modern collaboration and its impact on risk, insider risk and its effects, and explains these through real-world examples and scenarios examining cyberwarfare, government regulations, the evolution of risk management and compliance, and the role of data and privacy in managing risk. The aim of this part is to provide a comprehensive understanding of the various factors that contribute to risk in the modern workplace and how they can be effectively managed.

This part of the book contains the following chapters:

- *Chapter 6, The Human Risk in the Workplace*
- *Chapter 7, Modern Collaboration and Risk Amplification*
- *Chapter 8, Insider Risk and Impact*
- *Chapter 9, Real Examples and Scenarios*
- *Chapter 10, Cyberwarfare*
- *Chapter 11, An Introduction to Regulatory Risks*
- *Chapter 12, The Evolution of Risk and Compliance Management*
- *Chapter 13, The Role of Data and Privacy in Risk Management*

6

The Human Risk at the Workplace

Before we begin learning about different types of risks, let us look at an example to get a glimpse into how risks can look in real time. A computer engineer and a member of the IT team in the world's leading oil and gas company once opened an email and clicked a malicious link by mistake. The email looked authentic enough, and purportedly came from one of his colleagues, and the link that he clicked had an innocuous call to check a document that his colleague had shared. That opened the door to a group of hackers who proceeded to take over thousands of computers in the company network. The next few days at the company were frantic. This was expounded by the fact that the employee made this mistake during the holiday period, which meant that most of the other employees were not around. The mitigation efforts included taking the wires off all the impacted computers and turning most of the company's operations offline till the attack was dealt with. This meant that agreements were being faxed, printed, and signed on paper, and everything that needed to be officially emailed was being typed out on typewriters and sent by personal couriers or snail mail. The company's core production function was not impacted, but almost everything else was. The employee who made the mistake of clicking on a dubious link was perhaps a junior-level employee who did not mean any malice to the company. He did what he thought he had to do to finish his task. But his inadvertent mistake cost the company hugely both in dollars and reputation terms.

Modern enterprise (and by modern, I mean post-renaissance) has been built by and built around its core constituent, the employee. Capital (either as pure money, intellectual property, reputation, or a mix of all these) combined with labor (either physical, intellectual, emotional, or a mix of all these) has created stories of dreams. Invariably, all remarkably successful firms of the last decades and centuries have achieved success through a bunch of (visible or invisible) passionate, honest, and hardworking employees. These employees have been the soft assets that companies have tried to incentivize, retain, and train to get the most out of. These employees have also been, from time to time, responsible for some big goof-ups, accidents, and misdemeanors that have shocked and sometimes put the company in a squeeze, as we saw in the oil and gas company example in the previous paragraph. In the industrial era, the companies ahead of the curve identified the most crucial assembly lines and the most fearsome safety hazards and trained and monitored the employees handling them with abundant caution. Some learned it the hard way, some learned from others, but learn they all did.

As we aspire to restore the new world post-pandemic, as already established in the previous chapters, almost all companies are digital-first. And while assembly lines and industrial equipment remain top of mind in some sectors, the digital workplace goes beyond the boundary of sectors. The crucial thing is that with companies of all sizes and shapes now sharing almost all of their business processes and information and workflows digitally, the list of the most crucial employees handling the most sensitive equipment has expanded to *everyone*. Since we will expand into it, let us also refer to the seminal research from Eric D Shaw and Lynn F Fisher (you can read the paper here: `https://bit.ly/ShawFischer`), where they propose the following eight types of risky employees in a company, listed as follows:

- **Explorers**: Curious individuals who commit violations in the process of learning or exploring the system, mostly without malicious intent; they are unaware that their activities violate company information security policies (or such policies may not be in place).

- **Samaritans**: Individuals who bypass protocols and hack into a system to fix problems or accomplish assignments, believing their efforts to be more efficient than following approved procedures.

- **Hackers**: Individuals who have a prior history of hacking and continue penetrating systems after they are hired. These individuals have installed logic bombs or other devices in company systems to serve as job insurance when their activities are discovered. (They will defuse the trap in exchange for severance considerations.)

- **Machiavellians**: Individuals who engage in acts of sabotage, espionage, or other forms of malicious activities to advance their careers or other personal agendas. They include those who steal intellectual property to become consultants, those who sabotage competitors (or superiors), and those who cause outages to facilitate their own advancement or ability to gain attention. Machiavellians may also use their skills to advance social agendas.

- **Proprietors**: Individuals who act as if they "own" the systems they are entrusted with and will do anything to protect their control and power over this territory. They may actively resist threats to their control and are willing to destroy or damage the system rather than give up control.

- **Avengers**: Classic disgruntled employees who act impulsively out of revenge for perceived wrongs done to themselves.

- **Career thieves**: Individuals who take employment with a company solely to commit theft, fraud, embezzlement, or other illegal financial acts.

- **Moles**: Individuals who enter a company solely for the purpose of stealing trade secrets and other information assets for a competing company, outside group, or foreign country. While the research is more than a decade old, we have seen a few archetypes persist that we will refer to in this chapter. Please note that the list of archetypes includes the authors' own observations together with the archetypes that we feel persist from our own research. In this chapter, the main topics we'll cover are as follows:

- **Innocent intent** – We will ponder over this archetype in more detail than any other, because it is more widely prevalent than any other risky persona.

- **Good worker** – There are some employees who step in and save the day. And then there are employees who, in their infinite enthusiasm, step in to save the day when they should not. We will discover how they can harm a company.

- **Self-obsessed** – Those who take self-interest to a new level.

- **Rebel intent** – Those who cannot palate the perceived or real injustice done to them and turn to avenge themselves.

- **Malicious intent** – Those who can be moles who plant themselves or plain hucksters scheming and brewing plans till they can capture what they are seeking.

Innocent intent

The innocent employee can be ignorant of the dangers lurking around, as the following figure shows:

Figure 6.1 – A student unwittingly helping someone copy from their paper

An innocent indiscreet is, by prototype, a persona very omnipresent across all kinds of companies. You are, on average, more likely to be this persona than any other that follows. For the sake of this discussion, let us personify this innocent indiscreet persona and call her Ingrid. Ingrid is friendly, hardworking, yet affable, and has a sense of ownership over the company. She loves her company and her own work and has a good relationship with all her co-workers. Ingrid works for AGeneralExample Inc, a company that sells sports and collectible merchandise both through leading e-commerce and physical retail stores and through its own website. Ingrid is a designer, and her designs are a key factor in many of the top models that are sold by the company.

For us to categorize someone as passionate and dedicated a worker as an innocent indiscreet might sound diabolical to some of you. So let us now go through Ingrid's typical day of work and figure out how she has the propensity to be indiscreet and how that might expose both her and her company to various risks.

Start of day

Ingrid walked into the office and greeted (and was greeted by) her coworkers. Smiles were exchanged. She logged into her large terminal, with additional graphic capabilities to help her design work. She checked her mailbox, gave a look at her to-do list, and opened the most important work item. It was a design idea that she had to work on. She opened her design software UI, loaded the initial draft of her thoughts, and as the initial thoughts started forming in her mind, she stood up and walked toward the canteen for her first coffee of the day. She *did not lock her terminal while doing that*, and why would she have done that? She trusted her colleagues, and her workplace had biometric access control.

During the day

After having given a solid 2 hours of intense work to her design idea, Ingrid knew that lunchtime was approaching. She gave her wireframes and doodles one final flourish and saved them. While doing that, she made sure that she *saved a copy of that design in her personal cloud drive* as well. She did not do that for quite some time, but she learned her lesson hard, as suddenly, all work became remote after the pandemic one day. As she was stuck at her home and her company was trying to make remote access possible to her design terminal, it took a full week for things to finally work out. For that week, without her designs, she truly felt unemployed, a feeling she did not enjoy. Now, every time she worked on a design, she kept a copy in her personal folder and felt in control of her own work. She also knew that secondary personal storage would come in handy the next time she was on vacation and someone in her company called to ask (they always do, especially during vacation, don't they?) her to resend one of her works.

Wrapping up

Well, the afternoon meet-up with the product and business teams was chaotic, as usual. The business team wanted the designs to be simple enough for the masses, the product team had different expectations in terms of segmentation, while the web team wanted everything optimized for mobile. Ingrid took a deep sigh and looked at the final two versions of the design she had been able to get the whole project team to come together on. It was just that there was no consensus on the final output from those two. Ingrid took a lot of pride in her work and knew that the team also rated her so well that they would finally go with whatever she put her weight against. But Ingrid's pride came from her own pulse on people's choices and aesthetics. *She went to her company intranet, downloaded both her designs in low-resolution images on her mobile, and shared both with her small intimate group* of designer friends – "*A or B guys? Chocolates on me at the weekend get-together.*" They were her conscience keepers and more often than not, their suggestions came from an impartial, yet knowledgeable place. She logged off her terminal, packed her bag, and left for the day. The newly opened Italian boutique bakery on her way home implored her to come in.

A quick relook at the risks

Throughout Ingrid's day, we kept highlighting actions she took that might turn hugely risky for her or her company. Let us look at them and understand the risks better:

- **Leaving the terminal unlocked**: Leaving the terminal could cause the following risks:

  - **Risk of the Peeping Tom**: A vendor resource, a non-company visitor, or an unauthorized employee- any of them can glance a look at an unlocked terminal. This can expose proprietary design or other information to someone who should not be seeing it in the first place. For a lot of sensitive information, even a single view is damning enough as the person who stole that view could form full-blown ideas and thoughts, or fit a piece of a puzzle using that view. In a world where cameras have become ubiquitous through the cellphone, all it takes is a quick second for an open terminal to find its way to a competing company via a screenshot.

  - **Risk of unauthorized access**: Not only can an open terminal's visible content be screenshot and passed around, but the whole system can also be breached in a matter of seconds. Ingrid knows her colleagues, but she does not know whether one of them has been fired for disciplinary action or has got an unfavorable review and are on their way out within a few days. The risks of a system passing into the hands of someone with malintent are humongous.

- **Storing files on personal cloud storage**:

  It is extremely common for most company employees to make a backup of their corporate data on their personal clouds. They do it generally from the following places of misunderstanding:

  - **Ownership of corporate data**: While an employee creates a lot of information, IPs, and assets for the company, it is a common misconception among employees to feel ownership over that data. The fact is that they are critical custodians but not the owners of that IP or asset.

  - **Lack of awareness of consequential risk**: Personal cloud storage is more commonly breached than enterprise cloud storage. More importantly, personal cloud storage files live at the risk of not getting the careful treatment that they deserve. Think of an important IP containing a file that gets shared with a third party when Ingrid chooses to share a personal album/folder.

- **Downloading company data on personal, unmanaged devices and sharing socially**:

  While this instance of social sharing might look obviously problematic to most of you, if you are running a business where an employee can download a file on an unmanaged device, you have to think about that as a violation and a risk. Ingrid, in this instance, had nothing but good intent. She wanted a basic check and user research on the go to validate something that would have resulted in good business outcomes for the company. The problems are as follows:

- **Downloading intellectual property on an unsecured mobile device is risky**: Mobile devices are the new battlegrounds of cybercriminals and well-intentioned manufacturers. Recent research (`https://bit.ly/3h38uEA`) found a set of 400 flaws that can potentially turn a billion devices into spying tools. That impacts 40% of mobile devices around the world, 90% of which are in the US. This is not to say that the device makers, chip makers, and OS makers are not fighting that, or that they won't come up with a patch to fix these vulnerabilities. This is to give a snapshot of the threat landscape in the heavily fragmented mobile world where multiple versions of the OS, the motherboard, and the hardware are all existing simultaneously, giving the attackers a fertile ground to find one door or the other.

- **Sharing information socially opens the gate for intellectual property to be stolen or misused**: There is a world of social and chat applications that have made communication between people easier than ever in human history. However, when that ease meets corporate information, it makes for a disastrous recipe more often than not. Ingrid shared something corporate-critical with a bunch of people who are not in any legal or contractual relationship with her employer. Apart from multiplying the mobile threat nodes as mentioned in point 1 by the number of her friends, she also opened up the risk of someone mis-sharing or misappropriating the information further. The pandemic has increased the trend of social media usage during work hours and the risk of bringing sensitive information to a public network has increased proportionally. Take, for example, a workplace selfie, which might contain a whiteboard with sensitive business details in the background. Or the trend of sharing company ID card pictures on the day of joining or leaving a company – many times these ID cards contain an employee ID or some sensitive strings that provide a valuable pattern to a possible hacker lurking on the internet.

- **Apart from business risk, corporate information in the wrong hands can impact other co-workers negatively too**: In general, any corporate document has many stakeholders. Some whose business outcomes depend on it, some who have contributed materially to it (such as other designers in Ingrid's project team in this example), and some who have contributed ideas. There have been instances when people have come across some work they had done attributed to someone else on social channels and feel robbed of their efforts.

Let us now look at the second risky persona at the workplace in the next section.

Good worker

Some employees can be too helpful in ways that can hurt a company. The illustrative wall that the helpful figure in the following figure is helping a fellow climb can turn out to be a security breach as well:

Figure 6.2 – Did you check whether the person was supposed to clear the wall?

Every workplace has its own talismans. A super-employee who knows it all. A tech geek who you go to when you cannot find the solution to a tech impasse, as simple as making a productivity software or communication tool work. A person who knows what happens behind the scenes, who knows the organization tree and all cross-team linkages. For the sake of this example, we will call that person James. James was the product marketing lead for the same firm as Ingrid.

James was one of the oldest employees of the company and he took pride in showing his badge number to the new joiners. He had worked across various roles in the company and proved to be a great asset in all the teams that he had worked in. This had shaped him to become the eternal good Samaritan of the company. A go-to person who people reached out to when they had any issues. A person who knew almost everyone in the company and could fix almost any problem. Let us look at his machinations through the three sections of a typical workday.

Start of the day

James entered the office just in time for the first meeting of the day. It was the sales planning meeting, and the output of this meeting was super important for the leadership meeting later that day. The whole marketing and sales planning team was already in the meeting room. But they had a blocker. The most important member for this meeting – the **Management Information System (MIS)** lead – was missing. James, as usual, had his number, but before he could have dialed, they got a call on the company landline phone in the room; it was him. The MIS lead mentioned that he had an emergency at home, and he was forced to work from home that day. He had not logged on to the call because he was struggling to access his mailbox and even the corporate VPN did not seem to work. James realized that every passing minute was going to cost his team valuable prep time for the upcoming important leadership meeting. He asked his MIS guy for his personal email address and *forwarded all the planning Excel files to him on the personal address*. There should be no reason to lose any more time over it and the MIS lead promised that he would send the fresh forecast file basis all that within an hour.

During the day

James had a quick lunch after which he had to rally the troops around for yet another huddle before the LT meeting. As he stepped up to his desk to pick up his laptop, he saw a group of his colleagues congregating around his neighboring desk. A couple of his colleagues had stepped out of the office for lunch and on their way back, found a pen drive lying unattended near reception. The pen drive was brand new and carried their company's insignia. His colleagues were arguing about who it could possibly belong to and what should be done with it. The consensus seemed to be to return it to the IT desk and issue a notice but James, the good Samaritan that he was, offered a faster way to resolve it. He *plugged the stranded pen drive* into his laptop and went on to check for the files to do a quick forensic check to ascertain ownership.

Wrapping up

James ended his day on a high. The LT presentation had gone well, and his team was going out for a quick nightcap. There was only one glitch: one of his project mates was struggling to finish work as he needed a file from an intranet site that he did not seem to have access to. As the rest of the team was about to venture out without him, James chose to end his misery and *shared his credentials* for the colleague to quickly access the site. It worked. As they stepped out for the day, James could not help thinking how different the mood of his group would be without his contributions.

Let's take a quick relook at the risks:

1. **Mail on the personal ID**: The official data leaving the virtual perimeter of the company is a significantly risky event, as we saw in the previous example of Ingrid. We are increasingly seeing companies trying to protect their corporate data both at rest and in transit. In the post-COVID world, we have seen a huge increase in the number of data leak events, and in this case, if the file James shared contained sensitive data with some PII or confidential information, the outcome, if the vendor resource decided to reuse it in some illicit manner, would have been hugely consequential for the company.

2. **Plugging in a stranded pen drive**: This could have been a classic *USB drop attack* in which attackers leave a USB stranded expecting an internal employee to pick it up and plug it into their computer out of curiosity. This could lead to any of the following ways of breach:

   - **Malicious code**: The USB could have a malicious self-executing code that could breach the computer and propagate within the company network further from there

- **Social engineering**: The USB could have led James to a malicious website and tried to make him submit personal or company credentials for misuse

- **Spoofing a human interface device**: In a highly sophisticated attack case, the USB could have been programmed to spoof a keyboard-controlled device and could have been used by a hacker to take remote control of James's compute.

3. **Sharing credentials**: Violating access control could have exposed James's colleague to information he was not cleared to see. This can have grave consequences in the case of sensitive files or information that is gated for business reasons. Furthermore, it could have been leaked or misused by his colleague deliberately in many ways. And colleagues turning rogue is not something very uncommon in modern times, as we will see in the next sections; all it takes is a bad review or poor appraisal to turn a friendly colleague into someone you couldn't have recognized.

After having seen the first two personas in detail, we will look at the next few personas in brief. The first two personas are very universal and for the general reader of this book, those stories helped illustrate the risk inherent in mundane day-to-day activities. However, the next three personas are not as universal. Limited as they are, their impact can be significantly more, and hence it is important for all of us to have an understanding of their motivations and modus operandi.

Self-obsessed

This persona is defined by Shaw and Fischer (`https://bit.ly/ShawFischer`) as follows:

> *"Individuals who engage in acts of sabotage, espionage or other forms of malicious activities to advance their careers or other personal agendas. They include those who steal intellectual property to become consultants, those who sabotage competitors (or superiors) and those who cause outages to facilitate their own advancement or ability to gain attention. Machiavellians may also use their skills to advance social agendas."*

The impulse to maximize outcomes in any scenario has always been there among all human beings. With the advent of a mostly work-from-home, mostly online, mostly mobile social world, that impulse has gone through a multiplier. The content factory that social media is and the ease with which people can don personas without a physical presence spurs the Machiavellian self to be at its imaginative best.

Martin is a colleague of James and Ingrid from the preceding examples. He works in the IT team and has a parallel boutique website with a couple of friends from college, where they offer a variety of remote IT services to a select bunch of customers. Martin sees himself as a full-time entrepreneur a few years from now. To that end, he has to appear as the modern tech prima donna, and he has made sure that his social presence is shaping him toward that. He also understands that his current company is broad enough in its work to have multiple streams of business that capture lots of best practices that might come in handy when he starts his entrepreneurial journey. To that end, he has started accumulating all the information that he might have a use for in the not-so-distant future.

As Martin sets about his work, he, like any Machiavellian, poses three categories of risks to his company. The first is data spillage during accumulation. He can clearly store something personally that might not be of use to him but can harm his company hugely later. Inadvertently, he might capture details of a sensitive nature that can find their way out. The second is IP violation, where he might take proprietary information and put it to his own use. The third is a cultural loss to the company, where his example might demotivate other employees who put in a hard shift day in and day out without any perverse incentive in mind.

Rebel intent

To paraphrase the bard, "*Hell hath no fury as an employee whose leave request has just been rejected.*" On a more serious note, the pandemic has been the most severe shock to a large number of companies in the world. The kind of shock that many companies could not recover from, while many others will take years, if not decades, to come out of. Naturally, financial performances have plummeted for a lot of them. It leads to a roaring need for cost rationalization and eventually, it means downsizing employment or reducing paychecks.

Against the backdrop of this, we have to consider the kind of people who take negative outcomes to heart. A negative outcome impacting us is fairly common. Most of us find a way out of it either by understanding that we need to change something within us to keep growing or by making external changes such as getting out of poor work culture or moving away from a bad boss. But for a few, it becomes a matter of grave injustice and the need for retribution.

Looked at in conjunction, COVID has given rise to the number of perceived misgivings as well as negative outcomes. The avenging Adam is someone who can be the proverbial needle in the haystack when it comes to finding him but can hurt either the company directly or as a consequence of hurting the peers/superiors he takes out his ire on. A few known instances of risks that such employees can lead to are as follows:

1. **Sabotage**: From willfully destroying other people's work to deleting or misplacing crucial information, the range of misdemeanors is long. In the case of a large company, an employee who was terminated validly due to performance issues gained access to the finance system of the company and changed the bank records of a few vendors, as well as changed the payment terms and amounts. The company realized the financial implication of that months after the employee had left. In another instance, a coder played with the coding of multiple internal IT systems, leading them to dysfunction or malfunction. It led to months of chaos at the company and cost it millions of dollars in terms of man hours and opportunity cost.

2. **Impersonation**: This is one of the most common instances of revenge served. Most people let their guards down at work from time to time, such as Ingrid leaving her desktop on or James sharing his credentials. People jot down their credentials and leave them in sticky notes on their IT helpdesk in many companies. There have been cases in which disgruntled employees sent out malicious emails from the account of their superiors to a large number of partners and customers, leaving the company to mop up the legal and reputational consequences.

3. **Damaging reputation**: While both sabotage and impersonation can lead to reputational damage, an avenging Adam can be a great public spectacle as well. There have been instances of disgruntled employees leaving a company and using both real and fabricated pieces of evidence to inflict accusations on their ex-employer.

Having seen the fourth persona and the risks they bring to the workplace, we shall now move to the fifth and final persona in the next section.

Malicious intent

The fifth and final archetype of a risky persona is the most dangerous, if rarer than the others. In legal crime terminology, if the first two archetypes were innocent crimes and the next two were crimes of passion, this is the organized criminal who spends months and years plotting and planning their moves. There have been hackers and expert digital diggers who target a specific company, polish their resume, and put their best foot forward to get hired. Once inside, they dexterously study and inspect the digital and physical landscape of their workplace and plan meticulously. This can be a sole operator, hacking for their own benefit, or it can be a mole planted by a competitor.

In the post-pandemic world, a lot of data and workflows have moved online. While some institutions and companies have done that strategically, some have been forced to make tactical moves in terms of alternative workflows and tools. The clandestine operators thrive in such scenarios as it opens up new doors for them. Think of a crucial project that moved fully online. A lot of intellectual power of the project has now been reposited in one place and if that online location is not duly protected and monitored, the clandestine employee could exfiltrate all that they can in the blink of an eye. In brief, the clandestine archetype is known to posit the following types of risk to an institution/company:

- **Data exfiltration – big bang and leave**: The most common scenario is that of a clandestine employee targeting a specific project or the promising new launch of a company. They plan and execute to perfection each step of the way – creating a favorable profile, cultivating the right references, getting hired, and then while employed, drifting toward the project on target. Once they reach their target, they collect all the valuable information they need and resign and leave, only to remove all their traces. The exfiltrated data might be sitting with a competitor or even an upstart or a partner and the victim company finds out about this long after the event occurred.

- **Data exfiltration – slow but steady**: For all big bang leavers, there is an equal number of clandestine operators who might escape detection for longer. These are the people who exfiltrate information in small sizes over a course of months or even years. In a recent example, a gang of four employees forwarded sensitive emails to a common email alias outside the organization – almost a couple of emails daily over a course of months. They were caught as one of their forwards mistakenly got delivered to an employee who was not a part of their ring. But it is an exception that proves the norm – a huge number of such offenders are never caught. This is because it is too subtle, low in volume, and generally escapes detection even if companies have tools and solutions to watch over.

- **Source code and credential stealing**: With technology becoming central to many companies' business models, the incidents of tech theft have also increased. While a lot of companies and institutions are warding off attacks from the outside (via phish, spyware, trojans, or others), the clandestine employee can hurt them even more from the inside. In a famous recent example, an automobile company saw an ex-employee resign and join a competitor. They were surprised to see the competitor announce a product whose core tech was very comparable to what they had been working on.

The list of harm that a clandestine operator can do to a company is long. But the previous three bullets summarize a simple majority of popular examples from real life.

Summary

In this chapter, we saw the types of risks that a modern organization can be exposed to by its employees. We can categorize the personas of such employees into five buckets – the innocently indiscreet, the too-good Samaritan, the Machiavellian, the avenger, and the clandestine. The understanding of these personas will help you understand the type of risks that employees or participants can expose a company or an institution to. The demand of the modern age, especially in the post-pandemic transformed world, is for the leaders and shapers of these institutions and companies to proactively mitigate these risks.

In the next chapter, we shall go through the workplace from the perspective of modern collaboration. The knowledge of these personas in the workplace will be a good contextual cue for you as you go through that chapter.

7
Modern Collaboration and Risk Amplification

It was decades before the onset of COVID that modern institutions and corporations all realized the importance of cross-team collaboration to achieve big outcomes. The modern human culture still values the *hero* individually, but there is an increased awareness of large spheres of life and business where having 10 diverse brains looking at a problem together brings far more nuanced value than relying on one specific function or individual. The modern collaborative teams helped companies pool their talents and strengths and helped amalgamate creative, technical, and business forces across employees with diverse skills and backgrounds. It helped employees grow their skills by learning on the go from their diverse colleagues and shorten the critical path for their projects. As companies started breaking their functions or disciplinary siloes and creating outcome-focused virtual teams, technology became a crucial lever to help them facilitate collaboration. Remote working technologies and online collaboration tools and platforms not only helped the companies unleash their employees' talents but also helped the employees balance their work-life continuum.

But like all technological shifts, this enormous shift came with a point of caution, which was about securing the information, the tools, and the people who used it. The role of IT and InfoSec teams in large companies, and that of an admin in a small company, started becoming a balance of empowering their users while putting reasonable safeguards on the information. This forms the background for where this chapter will take you next. A very common opening of many traditional stories is *"Once upon a time..."* As this pandemic hopefully fades into the endemic stage, we will see a lot of companies and institutional anecdotes start with *"As the pandemic roared..."*

As the pandemic roared, almost all companies and institutions had to switch fully online. Before the pandemic, they were not at a similar stage of the digital technology adoption curve when it came to collaborative practices. After the pandemic, at least in the first few days, they were all expected to be at the same place in the technology adoption curve – to make sure their employees are fully functional from wherever they are. The partial or fully forced technological shift has ushered in a new era of modern collaboration.

In this chapter, we will cover the following topics:

- Evolving to the new workspace, and analyzing the journey to enable this transformation in the company
- Versatile collaboration – some questions that companies enabling it face
- Challenges of a hybrid setup
- Future devices and their risk profile
- Polarization of opinions and its possible impact on modern collaboration

Evolving to the new workspace, where the flow of information is versatile

For a wide range of companies, modern collaboration was already a partial reality. Even in the physical workspace, the tools and technologies to facilitate collaboration were in various states of adoption. To give you an example from 2019, right before the pandemic was to hit us, in IDC's US Enterprise Communications Survey, 2019 (`https://www.verizon.com/business/resources/articles/s/new-collaboration-tools-rolling-out-an-enterprise-application/`), one out of every two respondents said their organizations were currently using unified communications and collaboration, but 41% of employees in these organizations weren't using the tools available to them. The USA is one of the most technologically advanced markets and large enterprises are generally more receptive to tech adoption than others. In this segment, to summarize, half of the companies did not use collaboration tools and in the half that used them, only about 60% of their users actively used collaboration tools.

In the select subset of top corporations in the USA, only about a quarter of employees were using collaboration tools. One of the topmost reasons for that was the mismatch between what and how employees needed to collaborate and what they got in terms of capabilities. Later in this section, we will look at training and education separately. Outside of that, a whole lot of thought needs to go into the selection process for the right kind of collaboration tool. Depending on the profile, there is a whole host of software and services that facilitate specific needs of the employees – the needs of a development team might vary from the needs of an HR team, which might be very different again from, say, the finance team. The lack of uniformity in adoption leaves large gaps for organizations when it comes to security. Take, for example, organizations where they have commissioned WebEx as the primary collaboration tool. However, the developers, such as Slack, frequently connect on a freemium version of Slack. This creates a silo of the organizations' data that lies out of the view of their security team and is often referred to as **shadow IT**.

For a large chunk of companies, cost is a big factor. The known risk here is not only that of investing as per needs, but also that of identifying the tools' proclivity to facilitate collaboration safely. We have seen a lot of companies avoid making some investments pre-COVID and we have seen that employees find free versions of tech solutions to fill in the gaps left by that. The world of technology has exploded in terms of options, whether they are paid, free, or freemium. The selection dilemma is only compounded by the fact that a company not investing in a specific collaborative technology does not necessarily mean that employees won't use it. It only means that there is a risk of shadow technology infrastructure that can lurk invisible to the company's formal eyes. To clarify this further, we have created a graphic to give you a basic understanding of the expanse of the technological solutions landscape, which enables modern collaboration with or without a conscious choice having been made by the corporate leadership:

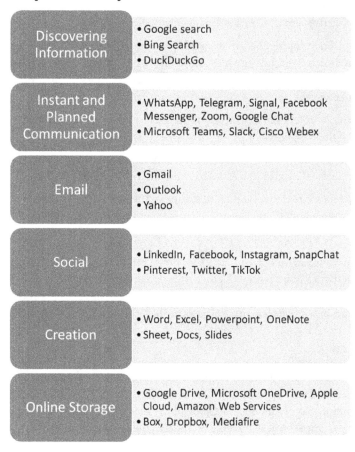

Figure 7.1 – List of technological solutions mapped to common needs

Let us delve deeper into this through a fictional example based on real customer scenarios. Let us assume that Rob is the owner of a medium-sized company (Imaginsta LLC) that manufactures and supplies garments to big brands. They have a team of frontline workers who do the handwork and manage production. They have the usual sales, finance, and HR functions manned by small teams. They also have a small developer team that manages their website and design repositories, and coordinates with their big customers to create workflows that parse the required design into small chunks of work and send directions to their production team seamlessly. Rob had encouraged his teams to be frugal with spending and maintained the rudimentary technology infrastructure to enable them. Before the pandemic hit, they had their full workforce working from their setup, which included both production and functional offices. Their typical flow of work looked like this:

1. The sales team managed relations with their large accounts. They interacted with them in person, through calls and emails. They maintained a pipeline of deals in a freemium CRM system. Once a deal was confirmed, they connected the marketing team of their clients with their own marketing and design/development team via email.

2. The design and development team iterated with the customers via email and finalized the required specifications and design after a few calls or, if required, in-person meetings. They received the final brief by email and created line-specific designs for their production team in the factory-like setup.

3. The production workers got a printout of the final designs and developed items under the supervision of their supervisors.

4. Once the product was delivered, their finance team tracked payments and updated records in basic spreadsheets. They also processed payrolls and tracked leaves using paid software.

5. The HR and finance team worked on desktops in the office and were able to use emails on their mobiles like all other employees. They also had ERP software that they used from their office machines.

As the pandemic hit, Rob had no option but to shut down production until the government allowed limited industrial activities to be resumed. Once they could resume partially, he had to allow most of his nonproduction staff to work from home. Rob invested in basic remote working technologies, including some cloud-based collaboration services for his office staff. But with time, he noticed the following points of concern:

- He had expected his production head to coordinate with design and development and get the design prints picked up. With time, the production floor team and design team started sharing designs on personal chat applications. On being asked, the production supervisor cited long-distance commutes to design houses as one of the reasons why they shared design print files on mobile apps. Once the designs came on mobile, the line and assembly workers also started asking for their pieces on WhatsApp. Rob realized that it could be very problematic, but he was confused about the best path forward. He had already invested in remote working capabilities for his office workers and felt that the cost of extending similar technology for his production staff would be prohibitive. He also felt that the production staff did not need all the remote working capabilities as other office workers.

- He saw similar fixes being used by HR and finance teams for all the gaps that emerged in the post-pandemic world. While his finance team was using a free web-based service for data analytics and dashboard, the HR team had started using a mix of popular consumer applications to deliver outreach to all employees.

- Rob used to run a tight ship from his office and was used to feeling in command of what was happening in his company. With remote work, he felt out of sorts, even after having invested heavily in some of the technologies. It seemed to him that the landscape of technology was infinite, and he kept running into holes that he did not even know existed.

As authors of this book and as cybersecurity professionals, we have interacted with a few "Robs" in our line of work and we have seen them address these issues in different ways. But we will try to cull out the universal themes from such cases. How the flexibility and versatility of modern collaboration impact organizations is mostly a net positive, but there are a few concerns that are thematic and need to be addressed. We have compiled those themes as commonly asked questions and tried to address them in the next section.

Versatile collaboration – some questions

We have worded themes as questions to represent the broad concern areas that businesses have to consider as they migrate to modern collaboration. Here are the top ones:

1. *It seems like a lot of investment. What is the business case and how do we justify it?*

    Starting from the business owner to the functional heads and the department heads, it is important to build an open conversation while evaluating the best ways to collaborate. The specter of *too much investment* is created either because the decisions are piecemeal and made to cater to different small groups or because the business benefits have not been seen completely. It is contingent on business leaders to do a holistic evaluation and build an organization-wide transparent conversation about technological choices.

2. *Who owns this? Should it be the CISO's business or CIO's or someone else?*

    Many of the technology choices have second and third-order impacts across the organization. The ability to work from anywhere has implications on the mental health of the employees and the culture of the workplace. HR as a function has come to the forefront in managing and envisaging the impact of this shift. Choice of tools and rhythms impacts all departments, including finance, operations, sales, and marketing. In current times, many leaders are trying to fix ownership for something as pervasive as collaboration and the obvious contenders are the IT teams. However, it is important to note that *every business is a technology business and technology is everybody's business*. While IT function can lead the conversation, the conversation must be owned equally by different functional and business heads.

3. *Will modern collaboration impact my workflows? How?*

    A change in technology entails a whole change in the ways of working. Leaders need to realize that the adoption of new technology impacts the flow of work at both the micro and macro levels. Take, for example, a business dashboard. In the old way of working, a **management information system** (**MIS**) managing resources published such business dashboards after taking oral and mail inputs from sales teams. With modern collaboration, all sales folks can be updating a common data repository and the data tool may surface insights automatically after some basic configuration. What happens to the MIS executive now? Depending on the needs of the company, the MIS executive can now be repurposed or retrained, though alternatively, they can elevate their work to high-level analytics beyond the tool. The important thing is to be mindful of the impact on workflows and gain the most out of modern collaboration.

Who is responsible for securing this? Say someone hijacks an online meeting, or an external user steals private data from a collaboration channel – who is accountable to ensure these breaches don't happen?

The accountability to lead this lies with the subset of the organization that is tasked with security and data protection. By "*lead*," we mean preparing a framework to secure these modern collaboration tools. But the execution of that framework is a shared task, the responsibility of which lies with each individual of the organization. Depending on the complexity of an organization's data landscape, the central security/data protection team can determine the minimum privilege that each user has in terms of scheduling or conducting a meeting, for example. But the end user must be aware of the responsibilities that come with that privilege. For example, let us say that a company has a policy of restricting access to the chat after the meeting for guest users. This ensures that a document or text that's shared during the meeting cannot be exfiltrated later by an external person. However, it remains the responsibility of the participants to not share any data that is not meant for that external person, even during the meeting.

4. *How do we think about culture when people don't meet face to face?*

There is a significant cultural impact of modern collaboration. A lot of coffee corner conversations, lunch gossip, and water cooler snippets tend to disappear with remote work. Companies that harness their culture look at modern collaboration as an opportunity to reinvigorate their culture. There is a need to relook at the tenets of your culture and reimagine them in this new reality. There is a lot of value that unstructured employee interaction brings to the table and there are a lot of ways in which many important cultural conversations and practices can find a new home in modern collaboration. Being intentional about this shift is key for many companies today.

5. *What does the future look like? Are we doing this to manage the current challenge or are there future consequences that we are not even thinking about?*

This is one of the most important questions pressing the top business leaders today. To address this in detail, we have devoted the next few sections of this chapter to answering it.

We will start by looking at the challenges of a hybrid reality.

Challenges of a hybrid setup

Hybrid work before the pandemic meant a mix of work from home and work from the office. However, the prolonged lockdown and forced work from home changed things profoundly for employees and companies alike. This is hybrid 2.0 and is vastly different from the pre-pandemic hybrid work model.

The first thing to note is that for a lot of young employees, the lockdown was a rude disconnection from their social habitats. Co-living and sharing accommodation with a larger group of friends increased as a trend among youngsters post-pandemic. Living together with a group was essentially the insurance policy against forced de-socialization. On the other end of the spectrum, a lot of companies either under revenue constraints or to cater to the shifted footprint of their employees moved from large hubs in central metropolitans and adopted co-working spaces across cities. Both these trends make the melting pot of modern collaboration a different beast altogether from the perspective of data security. Think about two friends co-living and working for competing companies. Think about how many times they can expose sensitive internal data to their peers. Or think about the trend of hundreds of photographs spread across tens of consumer apps and the probability of an important screengrab lying in the background ripe for exploitation by someone untrusted.

The second thing to note here is that a lot of people changed their locations almost permanently. The pandemic impacted cities and metropolitans more severely than smaller locations in many countries; the mortality and infection rates were both higher in metropolitans owing to a city having multiple points of exposure. Also, in some cases, there were losses of jobs or pay during the lockdowns. The ease of living in big cities declined during the lockdown and the appeal of small places increased. All these contributed to a large chunk of the workforce migrating to either their native places or small suburban areas. Caribbean beach towns and other popular destinations further captured this trend by incentivizing remote workers to move there. The result was that *work from home* was replaced by *work from anywhere*. The repercussions from a data security and enterprise assets protection perspective are immense. In *Chapter 2*, *Going Digital*, and *Chapter 3, Visible and Invisible Risks*, we touched upon these aspects. Work from anywhere pronounces those risks. Working from an idyllic beachside café makes for a great Instagram post. It also makes for the greatest nightmare for your information security leaders. Did the employee connect to the Wi-Fi of the beachside café? Was it encrypted and secure? In most cases, nobody knows. Sustained work from anywhere also meant that a lot of employee devices that were usually serviced and attended to by the company's IT desk risked missing some particular upgrades or patches. For business continuity, a lot of these patch management policies were relaxed and opened vulnerabilities for the organization. Devices in general are a separate topic of interest in modern collaboration. In the next section, we will delve deeper into that.

The third aspect is the collapsed physical boundaries in the hybrid world and the second-order effects of that on compliances and regulatory postures. The first-order effect of a hybrid setup is starting to be addressed – we see companies trying to mimic some offline customs online and we see them invest in empowering their employees to make the best-suited decisions to work from home or the office. The second-order effects of this are tougher to predict in regulatory terms. One recent example is of a banking company that had convened for a meeting of its board members where half of the attendees were in the office while the other half were logged in remotely. There was some glitch in the call due to which a couple of important members were not able to hear anything or speak. To solve the glitch, IT had to intervene. Since this was not an all-office setup, the IT person was added to the call so that he could investigate the issue online. As he probed the issue, he could see a lot of sensitive files in the chat of the call and could probably take screenshots or copies in the brief duration he was online. For the rest of the attendees of the call, the realization came as an afterthought that they had breached regulatory requirements by exposing extremely sensitive information to someone who did not have the credentials to see it. We saw the regulatory body release circulars to all **Banking and Financial Service Institutions (BFSI)** after this incident, but it is important to note that this was just one example of the second-order effects of hybrid work. There are possibly many other ways in which a hybrid setup can expose companies to regulatory tail risks and leaders need to be cognizant of that.

Future devices and their risk profile

The landscape of devices has evolved rapidly worldwide in the last few decades. From bulky mainframes to nifty personal computers to super sleek smartphones, we have seen a generational shift in devices in terms of size, capabilities, and compute power. One of the tenets of modern collaboration is the ability for an employee to have a work-life continuum and hence be in control of the balance. Personal devices are very important for employees and with work from home/anywhere, these personal devices are significant in their ubiquity. A few notable examples can be smartphones, fitness trackers, smart watches, smart TVs, e-readers, and digital assistants. The use of these devices poses a significant security risk to the company. Before we clarify this through an example, let us provide some context. IoT devices have been present in the corporate and institutional environment for a long time. Be it the printer at the coffee corner or the biometric door openers, internet-controlled devices outside of core computing have been present and created concerns from a security perspective from time to time. In the world of modern collaboration, those nodes have multiplied with the entry of personal devices, home connections, and their invisibility from the traditional perimeter-ed office. Therefore, we shall look at simplified examples to understand this better.

It is opportune to start by mentioning a search engine called SHODAN, which stands for **Sentient Hyper-Optimized Data Network**. It started as a hobby project by a young programmer named John Matherly who was just curious to find how much information he could gather about devices linked to the internet. What started as a hobby project, in about a decade, materialized as a powerful search engine that could capture details about 100 million plus devices, their exact location, and the software that controls them. A large chunk of these devices is **Internet of Things (IoT)** devices, including a large chunk of industrial control systems, routers, webcams, smart devices, and more. As people started using SHODAN, they realized that they could easily discover the IP addresses and locations of a wide variety of devices. It was reported that a bunch of hackers targeted some large industrial control systems and easily hacked them. The modus operandi was very simple. All these industrial control systems (take, for example, a city's water or waste management system) were old and supposed to lie before a brick wall. The connectivity to the internet came after these systems were deployed and a search engine like SHODAN raised the veils from their existence. Many of these control systems had default passwords that the hackers could read from their age-old instruction manuals available online.

The lesson is very clear: IoT and the spread of IoT devices is a trend that will persist. Modern collaboration is another trend that is well adopted and poised to explode. The security custodians will have to be conscious of the ever-increasing surface area of intersections – there are ways in which future devices can be a security risk to an enterprise. We have provided an illustrative example of how these devices can pose security risks here. Note that this is a simplified diagram and that we have avoided getting into details of countless other deep and technical ways in which personal devices can be vectors of risk:

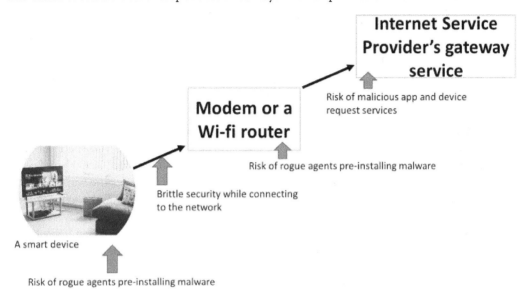

Figure 7.2 – Risk at every stage of smart devices connecting to the internet

The risk landscape emanating from futuristic devices such as IoT and others is a point of concern for many companies already. Now, let us look at another rising concern – the emergence of social media, the polarization of human opinion, and how it can impact modern collaboration.

Polarization of opinions and its possible impact on modern collaboration

The advent of social media and the constant flow of content from the firehose of the internet has already shown signs of immense impact on the politics and public commentary of our times. In the public sphere, we have seen unexpected events happening in the strongest democracies of the world, such as the raid on the Capitol in the USA after the 2020 presidential elections. While the world being full of different opinions and interpretations has been an age-old phenomenon, it is for the first time in the history of our civilization that those differences are so readily visible all around the clock to all the human beings on this planet. For many leaders and enterprises, the initial response to these angry debates was to ensure that the company stays out of political spheres as much as possible. It was soon very clear that in the era of modern collaboration and the young generations in the workplace, people will necessarily bring some of their political and socio-cultural personas to work. We have seen many start-ups and even larger companies understand that having no political opinion is not appreciated by the emerging new workforce, so some of them did two things in response (the causation might not be true for all of them; this is a conjecture of the author. The consequent implications stay valid irrespective). First, they encouraged employees to be authentic, true versions of their selves at work and encouraged them to freely express their opinions on various issues. Second, the companies themselves started asserting and expressing themselves on socio-cultural issues to ensure they appeal holistically to their customers. As an irony, both these stands have drawn their own fierce debates in the online world, where people are attacking or defending both these premises and some companies are further contributing to either side of the argument. Take, for example, Google. It continues to be one of the most attractive companies for talent worldwide and one of its charms was the transparency and the ability of all employees of all capabilities and persuasions to have free and open expression in their workplace. Things boiled over when an employee, James Damore, wrote a 10-page memo and circulated it to all internal employees (`https://whr.tn/3IKvF1E`). Without going into the merits of the memo or its critiques, it can certainly be said that the furor that the memo generated would have been a huge distraction for many Google employees. This included the top leadership, who had to spend hours in internal and PR meetings to handle the consequences responsibly. This brings us to the other kind of companies, such as Coinbase, whose CEO wrote a public blog discouraging employees from expressing opinions on broader societal issues at the workplace unless they had a direct impact on the company's core mission (`https://bit.ly/3tFJtGE`). The last words in both these kinds of debates have not been spoken, but it is increasingly clear that many companies will have to reimagine their future in the context of this polarization. The space for neutrality has receded already and may even be seen negatively. As companies and institutions, some might dislike being forced into taking positions on such open public debates, which might not have a direct correlation with their line of business, but it can be riskier to not take a side in these debates.

There is another significant consequence of this polarization and information overload that companies and institutions will have to deal with in the future. Employees working from the office or home will be in a consistent stream of information from personal, public, and corporate affairs. The initial posture of many companies in the past was to create a firewall to ensure employees could not access non-work-related websites or platforms during work hours. But this approach had its own limitations that were enhanced during the pandemic, which enforced a fully remote work regime. In the future, the companies will have to redesign and rethink their posture while considering this information blast as a given. To begin with, here are the important aspects to note for the future:

- **Polarization's direct impact on a company**: As more and more companies participate in the new knowledge economy, it can be expected that at least a fraction of employees will take diametrically opposite viewpoints on social, cultural, and political issues. These positions can be in direct conflict with the company's stated position or mission statement, and from a legal and governance standpoint, this can create complications for the company. Companies' codes of conduct have espoused universal values basis learnings that were considered civilizational. As some of those civilizational values are being stretched or challenged with various political and social lenses, the legislative landscape is also seeing a churn. It will be important for companies and institutions to be clued into the larger legal and political landscape and sensitize their steps or views in keeping with the larger picture.

- **Polarization's indirect impact on a company**: The overexposure to a myriad of opinions is also impacting the well-being and mental health of employees in general. The outcomes can be severe for companies. In the recent past, we have seen increased instances of employee burnout or even employees reacting strongly to non-work-related impulses in the workplace. This can lead to harassment claims, code of conduct violations, or even self-harm. In the traditional workplace, in-person interactions were the primary source for assessing employees' motivation, health, and well-being. In the era of modern collaboration, companies will have to develop technological muscle to keep in touch with some of these aspects. Sentiment analysis is one such example – we already see companies doing *social listening* to pick up the mood of their consumers/customers online. We predict that it will become equally important for companies to learn more about the sentiment of their employees.

That brings us to the conclusion of this section. Needless to say, this was a selective sample of a whole host of possible future risks but we believe that it is representative enough basis of what we know now.

Summary

In this chapter, we assessed the emergence of modern collaboration and its impact on companies going forward. At the very basic, companies need to reason out and carve their path to enable modern collaboration for their future needs. But equally, companies need to imagine the second-order impact of some of the modern collaboration trends. The physical reality of a hybrid workspace, the emergent landscape of smart devices, and the information overload with polarized public opinions all contribute to some of the challenges that companies will need to address while enabling modern collaboration. Seen together with the personas that we laid out in *Chapter 6, The Human Risk at the Workplace*, you can start to form a picture of how this can cause risk to companies from the inside. In the next chapter, we will try to evaluate these risks and discuss approaches in which we can do that.

8
Insider Risk and Impact

Insider risk can be defined as the risks a company, an institution, or a nation faces due to the actions that originate from the people within its environment. In *Chapter 6, The Human Risk at the Workplace,* we covered the personas of employees at the workplace and how some of these personas can lead to risky outcomes for a company or an institution. In *Chapter 7, Modern Collaboration and Risk Amplification,* we discussed how the shift in employee collaboration practices post-pandemic is opening up new vectors of risk for companies and institutions. We hope that the learnings from those two chapters have given you a good background for what we are going to discuss in this chapter. The topics we shall cover in this chapter are the following:

- A case study of insider risk in an anonymized real-world company

- Understanding the costs of insider risk – primary and secondary impact

- A summarized view of the overall impact of insider risk

Let us start with the case study first.

Case study – insider risk at Roposo Ltd

This case study is based on actual events that happened in a real-world company. We have anonymized the name to keep the focus on the learnings.

Context

Roposo limited is an iconic global multinational company with a long history of technological innovation and a diverse set of products and capabilities in fields as diverse as aviation and finance. This case study is related to the electric- and power-related parts of its portfolio. The company has a long history of dealing with power generation and distribution. The company manufactured all the key components of power plants and sold them to other companies around the world. Their expertise in manufacturing also translated into having a set of expert engineers and secret formulas and designs that could be used to tweak or tune existing power plants to maximize their performance and output.

They offered these design and optimization capabilities as a paid service to companies. Across the world, their experts and expertise were in high demand.

Actor and the plot

Patrick Dallas joined the company as a performance engineer in the first decade of this millennium and worked on some of their projects. He grew close to another employee, Serano, who left the company to start a business of his own. After having been employed for 7 years, Patrick went back to school. He completed a master's degree in business and returned to the same company and division. But it was within a year of his return that his company started noticing a pattern. Many of the projects that they bid for were bid for and matched by a new competitor. It was a large customer bid where they again lost out, which set alarm bells ringing. They probed the details and specifics. They found out that this competitor had outbid them handsomely but the price that the competitor had quoted was suspiciously similar to what they would derive as the base cost for this project. Basically, the competitor seemed to have a close awareness of their base costs and was bidding without any margin on top to establish itself as a business.

The crime

The company began investigating the credentials of this competitor. A very rudimentary web search showed them that the company was established by Serano, the employee who had left the firm quite some time ago. Further investigation showed that after joining back, Patrick Dallas had started downloading the most secretive data files from the company's servers and started exfiltrating them. As per public records of the case, he had downloaded about 8,000 documents that consisted not only of the *secret sauce* (some of the scientific and engineering proprietary formulae that lay at the core of the company's business offering) but also the marketing and customer details. On probing deeper and questioning Patrick, he resigned from the company. The competing firm was set up by Patrick and Serano. Patrick was passing intelligence related to trade secrets and a whole host of other information that would help them set up their company and get their first set of important clients. After being successful in exfiltrating the first set of crucial documents, he also approached and convinced an IT department employee to provide him with access to some other file stores in the company environment as well. Those stores contained super-classified intelligence that Patrick accessed and used. It was not very clear whether that was pure manipulation of the IT staff or whether there was some quid pro quo involved.

The aftermath

The case was reported to the investigative agency of the government. The agency spent almost 8 years tracking the forensics of the case. They obtained search warrants and scanned through their email accounts, servers, and cloud storage locations. As confirmed by the agency spokespersons, it was easily established that they had violated company policy and acted dishonestly. However, for this to be called a legal crime under the penal code, they had to establish the stealing of trade secrets with a high bar of proof. This took time and some luck. They were able to gather intelligence on the movements of

Serano (who lived outside their country's jurisdiction) and then arrest him while he had stopped over during travel in their country. The shock arrest also helped them to capture and examine his personal device and once they located the secrets on his personal device, it was easy to prove the crime to the court. Both Serano and Patrick were successfully charged and convicted.

The lessons

As you think about the details of this case and process some of the observations, a few points stand out:

- The time taken for the company Roposo Ltd to capture details of this case ran into months. It took the investigative agency years to make the case for prosecution. We will come to this again in a short while.

- The perpetrator was not only an insider, but he was also an old employee who had spent years in the company. If you look back at *Chapter 6*, *The Human Risk at the Workplace*, the main perpetrator fits the archetype of *self-obsessed* while the IT guy who helped him with undue access credentials might fit the archetype of *good worker*.

- Roposo had a data security regime in place but in hindsight, any security regime is only as strong as the weakest link. That is also because the governance and administration of these regimes is dispersed through multiple roles.

- There are quite a few unanswered questions that the Roposo team would have faced in the aftermath that are not publicly available but can be speculated. This case became visible because of a large deal and some striking similarities. What if there were other deals in the past that did not raise similar flags? How many such deals would they have lost? Was Patrick the only person in the organization who violated the policy or were there others? Was there any way for them to shorten the time it took to investigate?

- The involvement of the government investigative agency and the time taken to investigate and prosecute the crime were also telling. The legal rights of the employer are both extremely sensitive and subject to scrutiny, perhaps rightly so. But in such blatant cases, there is still a possibility for the odd bad sheep to manipulate the system and go scot-free. The most important thing from an investigative angle is to collect evidence and have a granular audit trail that captures the step-by-step deconstruction of any such activity.

- The impact of this case on Roposo Ltd was manifold. The whole exercise can be seen from the lens of lost revenue, the lens of employee hours spent chasing the case, or the lens of legal costs.

Continuing with the last point of this section, we will now delve into the various costs associated with an insider threat case.

Understanding the impact of insider risk

We will delve into some aspects of understanding the impact of insider risk in *Chapter 12, The Evolution of Risk and Compliance Management*, but our civilizations have progressed by understanding, forecasting, and mitigating risks better. Some risks are discrete-time events with well-defined outer and lower bounds of cost – take the example of regulatory audit risk. Certain companies in certain industries know that regulators would audit their books and practices once every 3, 6, or 12 months and if they fail that audit they will need to pay a fine of X, Y, or Z dollars. Some other risks are far more complex and continuous in their nature. In risk management, there is a popular saying that states that complex systems fail in complex ways. To estimate the cost of such risks, it is necessary to deconstruct the various parts of this complexity. A risky insider poses complex risks. The following diagram shows the whole gamut of actors, activities, and outcomes of an insider threat case:

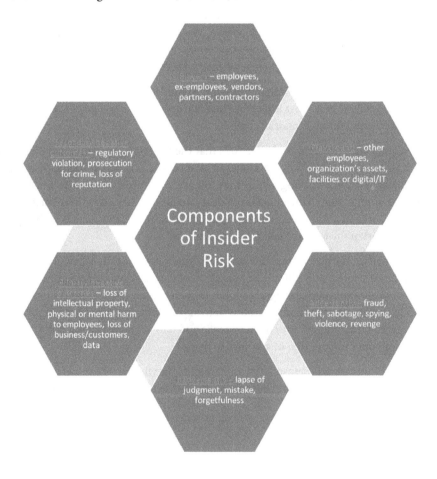

Figure 8.1 – The components of an insider risk case

As we can surmise from the diagram, the cost of insider risk is complex to arrive at because of the multiple components of an insider risk case. In the following sections, we will try to look at the components of insider risk from both primary and secondary impact perspectives. The primary impact factors are those that can be felt in immediate terms, such as loss of revenue, and hence are easy to quantify. The secondary impact factors though are more complex – things such as loss of reputation or the impact on employee morale cannot be easily assigned a dollar value. This is what makes the secondary impacts more intriguing. We will start with the secondary impacts first.

Secondary impacts of insider risk

As we enter the realm of analyzing various impacts associated with insider risk incidents, it is important to brush up on the fundamentals of risk assessment. Risk management is defined by the Oxford English dictionary as "*the forecasting and evaluation of financial risks together with the identification of procedures to avoid or minimize their impact.*" If you focus on the evaluation part of it, you will find a lot of art and science in there. Governments, regulatory bodies, and specialized independent institutions have created risk matrices and frameworks to help with the assessment and evaluation of risks. Let us take, for example, the **National Institute of Standards and Technology** (**NIST**), USA. NIST is a non-regulatory institution in the United States of America whose stated mission is to "*promote American innovation and industrial competitiveness.*" Among various other scientific initiatives, NIST also develops and issues directives and templates for security and privacy configurations (also termed *controls*) to help companies and institutions be compliant and mitigate cybersecurity and privacy risks. *NIST special publication 800-30* is one such template that many companies inside and outside the USA use. NIST SP 800-30 aligns with the risk management framework that helps comply with **Federal Information Processing Standards** for use in computer systems. In the following figure, we will see what the components of a risk assessment framework in a basic NIST model are. We will then pick the evaluative parts from it to form the baseline of our theoretical understanding before we delve into the secondary costs of insider risk cases. You can examine the details of the NIST model here: `https://bit.ly/3NXwwQw`.

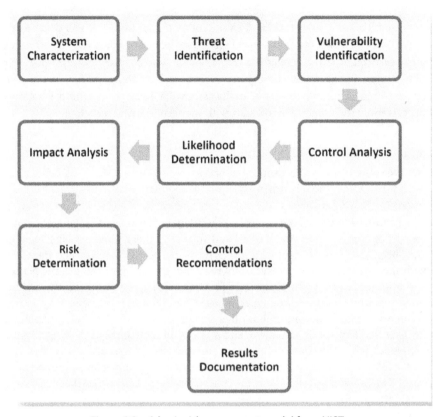

Figure 8.2 – A basic risk assessment model from NIST

At the center of this framework are two core steps – **likelihood determination** and **impact analysis**. The reference to NIST here is trivial in some senses. You can refer to multiple different risk management frameworks and at the core of risk evaluation, you will find two important questions – *what is the probability of this happening?* and *what would be the impact if this happens?* In some models, there is a third question too – *how precise is the risk measurement?* For this chapter, we will keep in mind the first two questions only, for simplicity. These two questions hold true for both primary and secondary risk evaluation but the difference is in execution. While we will talk at length about the various aspects of insider risk, we will expect the reader to ask these two questions for each of the scenarios for any industry, company, or scenario of their choice. The mental calculation is simple, like the expected value analysis in economics. Say there is an insider risk event that might happen with a probability of 0.001 (1 out of 1,000), and the impact in quantitative terms is 1 billion dollars. You can derive the cost of that risk to be 0.001 multiplied by a billion, which equals a million dollars. Let us now return to a few prominent secondary risk scenarios in insider cases.

The cost of reputation

One of the topmost costs associated with insider risk cases is the loss of reputation of a company or an institution. We are aware of high-profile harassment cases in big corporations or even non-government institutions. It is also common knowledge that many such cases never see the light of the day as the companies are under tremendous pressure and, in some cases, put the victims of such cases under similar pressure to hush up the case or try for out-of-court settlements. There are socio-political and gender issues at the end of such behavior and we would encourage readers to form their opinions, if not formed already, on those independently on the basis of abundant material available outside the book. To keep within the scope of this book, let us focus on the ramifications that such insider cases carry for the principal, which is the company or the institution itself. All such cases of insider risk that make it to the surface attract deep media, public, and regulatory scrutiny.

Take, for example, a renowned American airline that made headlines after its ground staff forcefully evicted a passenger from an overbooked cabin. There might have been institutional failures as well, but in public it was a live example of a company's insider(s) behaving in a manner that causes loss to the company. A classic insider risk case. The episode generated a long cycle of negative coverage in both traditional and social media. This is another thing that changed even more rapidly post-Covid; let us tell you how. **Online reputation management** (**ORM**) is a growing expertise with dedicated digital marketing or online reputation management firms managing the social media persona of a brand or a company. Traditional media management was a well-evolved art and companies had evolved practices to minimize the impact of the negative news cycle. ORM is a nascent field, despite the rapid advances made in the short span of time it has existed for. It came into existence with the social media explosion in the last decade or two. However, the nature of the internet is far fickler compared to traditional media, hence ORM is effective up to some extent in understanding and influencing the behavior of your social media audience. But cases like that of the airline touch upon the raw nerves of the social audience, and as the adage goes, "*The internet forgets nothing.*" Memes, GIFs, short videos, and various kinds of media together create a museum of reputational gallows that stays around almost forever. Post-COVID, we have not only seen social media adoption grow but also the time spent on social media has grown exponentially. As a result, the reputational risk for the odd risky case has multiplied.

So, how do we derive the cost of reputation? How do we measure the exact amount of reputation that a company lost because of an insider? Well, it is tough to put a number against it, which is why this is such a prominent entry in the secondary impacts section. But then, it is not fuzzy either. In the case of the aforementioned airline, their market capitalization decreased by $1 billion in a few days on the stock market. In the case of many other insider risk cases, where consumer data has been breached by an employee or personal data has been leaked or similar incidents have happened, the backlash in terms of stock price movement and public perception has been instant. Clearly, stock markets factor in reputation while pricing a stock. A good proxy for reputation is *brand value*. There have been many brand-value surveys where analysts have estimated that great brands have residual value over and above their net financial value, which can be expressed in dollar terms and, in some cases, is up to 40% of its net financial valuation. There also is an up-and-coming practice of reputation dividend valuation; you can check `https://bit.ly/38xhBMB` for an example.

To conclude, loss of reputation is one of the most prominent losses coming out of insider risk cases and risk managers will need to factor this in both primary and secondary terms for their specific sectors and industries' basis relative indexes and similar precedents.

The cost of employee well-being

While loss of reputation is market-facing and rigorously investigated, if not quantifiably understood, the loss caused by an insider risk case in terms of employee well-being and the consequences are less understood, and perhaps less deliberated. This is also an aspect that needs to be analyzed more in a post-pandemic world.

The International Labor Organization defined employee well-being as all aspects of working life, from the quality and safety of the (physical) environment, to how workers feel about their work, their working environment, the climate at work, and work organization. The reason why we put brackets around *physical* in the previous quote is obvious. You can substitute it with *digital* for the post-pandemic world. But employee well-being is essential for any company or institution to achieve its goals. There are at least three dimensions of employee well-being that can be impacted by insider risk cases and it is tough to put a numeric value against those impacts – social, emotional, and environmental.

Social wellness is about connecting with colleagues at your workplace. A workplace is essentially a social construct and the richness and openness of human interactions there play a critical role in the performance of the company that has created it. An insider risk case hits at the foundation of that social construct, which is trust. In our lived experiences, we have seen the odd employee's damaging behavior drive a wedge of suspicion in the sense of the collective. Think of a company that reported a treasure trove of sensitive customer data stolen by a fraudulent employee, or a company where a senior employee was found to be harassing a subordinate. In both cases, the rest of the workforce is thrown off balance with grapevines, rumors, and their own brains conjuring and creating a picture of everything that could be wrong with everyone else. The biggest harm is done to *trust*. Some companies have crossed such moments in their history and emerged stronger but even for them, there is a huge loss in the interim, a loss that can perhaps not be accurately quantified.

Emotional wellness is an individual trait. As we have mentioned in *Chapter 7, Modern Collaboration and Risk Amplification*, the individual of the modern world is struggling to adjust to the polarized landscape, among many other challenges. An insider risk case on its own would be a mental challenge to adjust to, for very stable employees, but most people are already on edge. These are the edges created by a mix of personal, societal, and political insufficiencies. In a post-pandemic world, an insider risk within a company will possibly cause elevated emotional turbulence to the average employee. That turbulence can impact their well-being and as a result the overall workplace mood for days or months. Not only is it tough to quantify in terms of productivity, but it is perhaps also inhuman. Because all corporate entities exist to serve a mission, however quantifiable those might be, the journey to those missions becomes severely impaired when the core constituents of that journey are in a state of turbulence.

Environmental wellness is the overarching layer above individual and social layers. It is the result of the interactions between these layers in some senses. A worried individual interacting with a tentative social milieu can very well lead to a workplace environment that is termed as toxic. In the case of an insider risk case, it is easy to understand through a hypothetical example. A company sees a massive insider violation and as a first step, as is known in many cases, suspends some of the privileges of all employees till the damage is contained. These suspensions are relayed over by a team of middle managers who do it in their own individual ways. The average employee sees and interprets this in their own way and draws her or his own inferences. Depending on the company, this could be temporary toxicity or something that could impact it long term. The costs must be paid though.

Primary costs of insider risk

To reiterate from the start of this section, we consider insider risk cases to be complex in terms of their ramifications. To understand the cumulative cost of an insider risk case, we are deconstructing them and looking at them through different lenses. We looked at the factors of secondary impacts in the previous section, while in this section we will look at some of the pieces that can be assigned a dollar value and hence are the factors of primary impact. We will look at the cumulative impact possible from all these pieces put together in the last section. Let us start with some of the prominent quantifiable losses that we have seen companies incur in the case of an insider risk case.

The cost of data lost

One of the prevalent forms of insider risk is the accidental or intentional deletion or removal of business-critical data. There is an industry-wide acceptance of the fact that every company is a data company today. A recent survey by New Vantage Partners, `https://on.tcs.com/3v46Itj`, highlights this. The survey is responded to predominantly by C-level executives of the topmost companies in the world. Among many other things, the survey highlighted the following:

1. 97% of the respondents confirmed that they were investing in data-led initiatives to become nimbler and data-driven.

2. Close to 80% of the respondents expressed fears that data-driven competitors will disrupt their industry and business.

We will not digress into big data and data analytics here but rather focus on the two primary classes of data – structured and unstructured. Structured data is data that conforms to a data model, is in a standardized format, and is found in schemas such as databases. Take, for example, an e-commerce site that sells thousands of **stock-keeping units** (**SKUs**) of different brands. The information would be cataloged in a standard database and would be an example of structured data. Unstructured data on the other hand is data that does not follow any preset model or schema. The most prominent example of unstructured data is text and media files. The same e-commerce site from our previous example would have unstructured data in the form of agreements with vendors and invoices, and some of that unstructured data might also be intellectual property or confidential information.

When an employee accidentally or intentionally deletes data or does something that irreversibly corrupts data, the resultant loss can be a monetary cost. If structured data such as catalogs are lost, the cost of that loss can be accounted for in terms of the cost of hiring manpower who will recreate those catalogs and the man-hours spent. If structured data such as a customer information database is lost, it has three costs associated with it. First, the cost of regulatory action, which we will touch on in the next sections, and second, the cost of recreating or buying the same information. The third cost is in those cases where an employee steals and wipes a company's confidential data, be it customer details or proprietary information, and causes business loss to the company, which we come to later in this section. In general, the costs keep increasing with the complexity and criticality of data.

The cost of damage to equipment, assets, and IT systems

The English word *sabotage* means a deliberate action aimed at weakening a polity, effort, or organization through subversion, obstruction, disruption, or destruction. A popular but unverified story explains the origins of the word thus – *sabot* is a French word that means "*wooden shoes.*" In 16th century Belgium, some textile workers who were disgruntled with their employer are said to have thrown their wooden shoes in the textile machinery wheels. The stuck shoes would damage the equipment and bring production to a standstill. Another famous reference is to the Luddites in 19th century England, who were a secret organization of workers that damaged textile mills to protest the increasing mechanization of the mills, which threatened the employment prospects of the existing workers.

Historical footnotes aside, an employee who perceives injustice at the workplace or has any nefarious agenda is likely to try and damage the assets of the workplace surreptitiously. With the rapid digitization of the enterprise landscape, the modern-day sabot is likely to be digitally thrown while the physical premises remain vulnerable as well.

A large educational institution parted ways with an IT employee because the employee did not agree to relocate to the suggested location of work. The employee believed that he had made his location preferences clear at the time of the appointment itself and suspected that he was singled out because of a racial discrimination claim he had filed against his superiors a while ago. He was given a severance package, but the employee carried a grudge against the perceived or real injustice of the whole scenario. As a way of revenge, he changed the password of his institution's cloud service provider account before leaving. The moment he left, thousands of users of this institution lost access to their emails and other services. The whole institution's daily work came to a standstill for days and when they tried contacting this now ex-employee, he feigned innocence. Apparently, the ex-employee then sought hundreds of thousands of dollars to restore access for his ex-employer.

The damage to a company's access and IT systems can be costly. The method to put a numerical dollar against this cost varies based on the actual scenario. In the preceding example, the loss of productive time for users can be costed by deciding upon a financial value that a productive user generates or is compensated for over that period. Any additional expense incurred to correct the system either in man-hours or extra money spent on solutions is added to this cost.

Direct monetary losses

From a costing perspective, we can club a few of the losses stemming from errant employees into the category of direct monetary losses. These are the easiest to cost as the loss happens in direct monetary terms. These are the cases where an insider took out money directly from the company by cheating the system or did something that caused the company a direct loss of business or deal opportunity. In the *Case study – insider risk at Roposo Limited* section of this chapter, we have provided an anonymized story of a power sector company whose trade secrets were stolen and used by an ex-employee to establish a competing business. The actions of that employee resulted in the company losing large deals and the cumulative amount of the loss was very clear in terms of the foregone revenue. In the *The cost of data loss* section, we touched upon the cases where insiders steal a company's customer records or proprietary secrets and either sell to a competitor or use them to set up their own business. The cost of these losses can be ongoing in nature if the case is not caught.

Similarly, there are cases galore where an insider misuses access or uses forgery on their way to directly swindling a company's money. Such instances can come in many ways – padding expenses, double dipping, fake vendor payments, and collecting kickbacks being some of them.

The cost of investigations

A significant amount of money is spent by companies in the aftermath of an insider risk case to investigate these incidents. Referring to the case study at the beginning of this chapter, the power engineering company took months to figure out the exact sequence of events. These months are inundated by productive security and compliance resources toiling and tracing the sequence of events from event and audit logs. Most modern corporations have established some form of a security program or a secure operations center to ensure the safety of their digital IT estate. These are manned by employees who are competent in various areas such as monitoring, incident response, investigation, escalation, and remediation of threats.

By industry estimates and studies, on average it takes about 200 days to detect a breach and about 80 days to contain it, though it can vary for specific companies. The level of sophistication of the insider's modus operandi determines how easy or tough it is for the investigators to get to the bottom of a reported or observed event. After the exact sequence of events leading to the breach or insider activity is investigated, further time is spent by precious resources on mitigating the gaps for the future and creating documentation for reporting. As we will see in the next section, the cost of investigation for companies can run into hundreds of thousands of dollars.

The cost of regulatory violations

You will read in *Chapter 11, An Introduction to Regulatory Risks*, about how governments around the world are legislating laws and regulations to ensure the safety of citizens' data and rights. To ensure compliance, they have fixed hefty fines on companies in case those regulations are violated. These fines have a clear value and in many insider cases, the company finds itself in violation of multiple regulations. For example, a risky insider can exfiltrate a European citizen's data and the company might be staring down multi-million dollar fines for GDPR violation.

A summarized view of the impact of insider risk

The 2020 business benchmark Global Report on *The Cost of Insider Threats* (conducted by the Ponemon Institute: `https://bit.ly/3jnK2Pn`) found out that across 204 organizations/associations, careless workers or project workers (which are also called contractors) were the underlying driver (the root cause) of 2,962 revealed insider risks/threats occurrences out of 4,716 occurrences (cases). Carelessness or negligence happened 63% of the time in many organizations/associations. A certification robbery (credential theft) happened almost 23% of the time, and a criminal insider was involved 14% of the time. The general expense of an insider risk occurrence cost overall 11.45 million dollars. Carelessness and negligence cost organizations around 4.58 million dollars and criminal insiders cost organizations/associations 4.08 million dollars. Accreditation burglary (credential theft) was likewise one of the top three expenses of insider attacks at 2.79 million dollars. Credential theft remediation itself costs organizations/associations 871,686 dollars.

Insider breaches caused by employees or leaders of an organization or within an organization are the deadliest and costliest of all, and also the hardest to detect of all the data breaches that happen inside an organization. 66% of all-out information records bargained in 2017 were the consequence of coincidental insiders, and as per the *2018 IBM X-Force Threat Intelligence Index*, insider risks/threats are the reason for almost 60% of all the cyber-attacks inside an organization. In the interim, misconfigured cloud workers (cloud engineers) and organized reinforcement episodes brought about by representative carelessness uncovered more than 2 billion records last year. While associations center huge assets around the alleviation of external dangerous elements, insider chances are probably going to represent a significantly more prominent monetary danger to the venture.

Insider fraud makes up almost 5 to 6 percent of the total cost of insider threats/risks. Insider extortion (insider fraud) is a kind of risk/threat that comes from within – a current or previous worker, a worker for hire, or a colleague can complete a fake plan that exploits the information or cycles they approach with regard to their work. These insiders frequently have mostly one-of-a-kind chances as admittance to significant information or assignments such as handling installments. A few occasions of insider extortion are perpetrated by people with criminal aims, yet insider dangers can likewise be the aftereffect of a human mistake or carelessness.

Following are some points that every organization or association should be aware of regarding the cost of insider threats/risks (all costs based on the 2020 study):

- Every episode or case including a careless representative or worker for hire (contractor) can cost an association an average of $307,111.

- Absolute expenses can race to $4.6 million every year. On the off chance that an occurrence or case of insider risk includes fraud or a hoodlum, the normal expense generally significantly increases to $871,686.

- Since 2018, the normal number of occurrences including worker or project worker (contractor) carelessness has expanded from 13.2 to 14.5% per association.

- The burglary (theft) of favored clients' accreditations or other information, in 14% of episodes, costs every association (organization) an average of $2.8 million yearly.

- The normal number of certification burglary (credibility fraud) occurrences has moved from one for every association to 2.7 per association.

- Criminal and noxious insider risks/threats, 23% of general episodes or cases, cost associations or organizations an average of $755,760 per occurrence and generally $4.1 million yearly.

- 60% of associations had more than 30 episodes each year. An examination of an episode is the fastest growing, which averages at around $103,798.

- Enormous organizations with 25,001 to 75,000 workers spent an average of $18 million over the previous years to determine insider-related episodes and cases, and even then, the cases are still on the rise.

- More modestly sized associations with a headcount under 500 spent an average of $7.68 million.

- The quickest developing ventures for insider risks were retail (19% increment per annum) and monetary administration (10% increment per annum).

The cost impact of insider risk runs into millions of dollars, evidently. We saw a lot of numbers in this section, primarily based on the Ponemon study, but the important thing to note here is that these costs are not unidimensional – companies spend this money together with a lot of focus and employee time. It is a distraction and all companies that have gone through a similar experience would probably want to avoid such episodes entirely.

Summary

In this chapter, we looked at insider risk and tried to decipher ways in which we could price these risks. An insider risk case can be complex to assess and hence we looked at the NIST model to see how risk managers try to assign a value to risks by trying to figure out the probability of the risk happening and the quantum of the impact associated with that risk. We then broke these insider risk cases and their impacts into two categories – primary impacts and secondary impacts. We looked at various components of losses arising out of insider risk cases and finally looked at the Ponemon Institute's study, which produced some trends and dollar values by studying many companies where insider risk activities had happened. While we have seen many examples of insider risk cases and a case study in this chapter, we will dedicate the next chapter to some elaborate case studies to give you a ringside view of such cases in the real world.

9

Real Examples and Scenarios

The Guinness World Record for the largest bank robbery is held by Dar Es Salaam Investment Bank. It took place in Baghdad in July 2007 and the robbers made off with about $282 million in cash. It was widely reported to be an inside job, with two or three internal security guards suspected of being involved in the robbery. In hindsight, it seems apparent that only an insider with precise information could have planned such a high-value heist. But with foresight, what level of damage can such insiders inflict on a digitized banking environment? What, in your view, are the common ways in which a rogue insider can target a bank?

"All happy families are alike; each unhappy family is unhappy in its own way."

- Leo Tolstoy

We can extend the message from the preceding quote to risks that an organization is exposed to and try to understand how one scenario can be completely different from another. A risky employee is risky for a company or institution, but the elements of risk, impact, and ways in which they can hurt the institution can be very different. The financial services industry has different vectors of risk from inside activity compared to a public sector institution, which has, in turn, a very different risk profile compared to, say, a consumer goods company. In this chapter, we will delve into select vertical markets and look at one top insider risk story from each of them. Here is a brief overview of the things we will delve into:

- A brief recap on the definition of an insider risk
- Insider risk categories
- Insider risk trends to look out for
- Four case studies

Insider risk – definition and threat vectors

In the world of cybersecurity, an **insider risk** or **insider threat** can be defined as the risk or danger arising from a **trusted insider** who may, intentionally or unintentionally, compromise the confidentiality, availability, and/or integrity of enterprise systems, data, and resources/intellectual property.

Personnel may knowingly or unknowingly expose the sensitive data and information of an organization to the external world while performing their normal tasks. This can result in a loss of reputation or a loss of high-value data, as well as creating a hole in the organization's network that goes unnoticed.

Insider threat has four main categories:

- **Fraud**: Employees, vendors, or clients who have access to an organization's internal assets/resources can steal, destroy, or modify critical data, devices, or systems for the purpose of personal gain or deception. Examples include phishing emails, business compromise emails, and identity theft, leading to unauthorized access to sensitive information.

- **Sabotage**: Insiders will leverage their legitimate access and ability to transfer data, with the intention of damaging the reputation of an organization. Examples include causing damage to computer equipment and networks deliberately, infecting a website with malicious malware, and causing a national power grid to shut down.

- **Intellectual property theft**: Insiders will steal a company's intellectual property (recipes and trade secrets), often for resale or to elevate them to a new position. Examples include stealing and leaking secret vaccination formulas and stealing new electric vehicle designs.

- **Espionage**: Insiders will steal information for another organization, such as a competitor, government, or nation-state actor. Examples include cyber-spies trying to gain access to sensitive resources, such as military information, data and activities related to research and development, a list of key customers, and payment methods.

Insider risk – behaviors and technical trends to look out for

Defense in depth focuses on the identification of certain behaviors that can be indicators of threats (intentional or unintentional). Some of these behaviors and indicators are covered next.

Behavioral indicators

Behavioral indicators can be identified and noticed based on the following scenarios:

- Attempts to circumvent physical and logical information security controls
- Routine risk exception requests for enterprise policy violations or training
- Displaying resentment toward co-workers, partners, and clients
- Unapproved and emergency leave away from work
- Being silent and quiet at work
- Dissatisfied or disgruntled employees, contractors, vendors, or partners

Technical indicators

Technical indicators can be tracked from the number of logs and alerts within an organization's security operation center or by an IT team, by analyzing the number of patterns from the logs and alerts, such as the following:

- An abnormal login time on a network and systems, such as logging in during off-work hours

- An increased volume of network traffic, such as a heavy download and upload of files

- Attempting to and accessing systems and applications beyond an employee's access privilege and job profile

- Sharing a large volume of data via external email addresses

- A high volume of printed papers and files

- Increased usage of multi-hop proxy such as a Tor (proxy) network to hide actual network locations

- Trying to attempt an entry in an unauthorized secure zone in an office, such as a lab, file archive area, or server room

- Using unauthorized devices to access enterprise resources or emailing sensitive information outside the enterprise

Using the MITRE ATT&CK framework to detect insider threat and behavior

The MITRE ATT&CK framework (`https://attack.mitre.org/`) is not only a model but also a published knowledge base of various **tactics, techniques, and procedures** (**TTPs**) used in a malicious attack or by an insider. Tactics are general steps while techniques are more specific actions taken by an attacker.

While the focus within this framework is on an external attack, these TTPs can also provide valuable guidance to a security analyst in understanding patterns and data sources. For example, in the event of an insider threat or attack, the logs from Windows Active Directory authentication can provide lots of insights, in terms of the login dates and times and the system accessed by the end user.

Here is one such example from the MITRE ATT&CK Framework Tactics for Data Exfiltration: `https://attack.mitre.org/tactics/TA0010/`.

As per this framework, **exfiltration** is defined as an adversary trying to steal data.

Exfiltration consists of techniques that adversaries may use to steal data from your network. Once they've collected data, adversaries often package it to avoid detection while removing it. This can include compression and encryption. Techniques for getting data out of a target network typically include transferring it over their command-and-control channel or an alternate channel and putting size limits on the transmission.

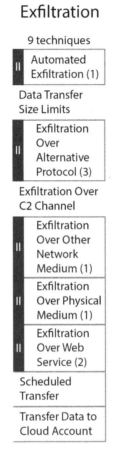

Figure 9.1 – The 9 techniques in exfiltration as per the MITRE ATT&CK framework

In the preceding diagram, we can see that there are nine different techniques under the exfiltration category; details can be found on this web page: `https://attack.mitre.org/tactics/TA0010/`.

Case study 1 – exploits in the life sciences sector

My team from my previous role in a Big Four consulting firm worked very closely with an information security office in early 2019 for a Fortune 500 life sciences company, investigating and providing a **root cause analysis (RCA)** of confidential and sensitive data theft by one of their vendors.

We were designing a centralized risk register for one of the most sophisticated global life sciences corporations in western Europe. Suddenly, we were called into the chief security officer's conference room, which had representation from legal, the chief data officer, marketing, investigation, and fraud officers, the chief research officer, and the chief launch officer (launching new products on the market).

One of the most critical formulae from their animal disease control department had been showing up on the dark web for the last 8 hours, and no one was aware of the theft until the chief security officer received a call for ransomware. This is a form of malware that encrypts a victim's files. The attacker then demands a ransom from the victim to restore access to the data upon payment. Users are shown a notification on how to pay a fee to get the decryption key. Typically, the ransom amount is demanded in any of the cryptocurrency wallet addresses and ranges from a few hundred dollars to thousands of dollars in value.

The organization decided against paying the ransomware after consulting with the ethics and legal departments and the board of directors, and it decided to start taking its core resources offline to contain the damage. The organization's CEO was concerned about the risk of exposure and brand damage. The cyber investigation team with law enforcement, the Cyber Forensics Council, and the Risk and Compliance Office tried to quantify the damage to the brand by the data theft and, at the same time, perform an **RCA**. The replication of the exploit scenario with multiple inputs, such as user accounts, reviewing event logs, and structured and unstructured data transfers, was very time-consuming, technically challenging, and involved some complex technology and processes.

After 22 days of investigation and a replay of events, the identified root cause was found to be an insider (a third-party vendor) authorized to access sensitive data via a remote channel. The insider clicked on malware that was sent via a phishing source, in turn compromising the access channels to the entire master data directory. Insider risks can occur with highly intentional threat vectors, who will steal data or intellectual property for competitive benefits, brand damage, or disruption to an existing business.

Solution and outcome

After this incident, the organization invested heavily to start its zero trust journey (trust but verify) for their entire network. Tabletop exercises for vendors and monthly phishing scenarios were rolled out to the entire business unit.

The incident was an eye-opening experience for the board of directors, audit committee, external auditors, senior leadership, and shareholders. The board of directors was committed to improving cybersecurity and risk management resiliency plans. External auditors and shareholders were provided with regular updates on the cybersecurity program.

Case study 2 – a victim of a phishing attack

During 2020 and 2021, we saw an almost 300% increase in phishing attacks according to the Microsoft Digital Defense Report (`https://aka.ms/MDDR`). Phishing emails are types of emails that a malicious attacker crafts specifically with a malicious link. When an end user clicks on it, malicious code can be downloaded onto their device, or the link will redirect them to a fake website that may ask them for their sensitive banking details or username and password.

One such example involves an email that was sent to one of the senior members of a reputed university in Australia. This staff member clicked on a malicious link, which resulted in almost 700 MB of data being leaked. This may not look like a huge dataset from the numbers; however, the information that was leaked included sensitive information about the staff as well as students, such as names, addresses, phone numbers, dates of birth, emergency contact numbers, and tax file numbers. Based on this information, bank details and students' academic records were leaked and compromised.

Solution and outcome

After this incident, the company invested in security awareness and phishing simulation services and implemented a better technical solution for anti-phishing and email filtering. They also invested in **data leakage protection (DLP)** and **cloud access security broker (CASB)** services.

The security team started using phishing simulation results as a part of its management reporting. This **key performance indicator (KPI)** helped them to understand the current risk exposure, due to a lack of security awareness and a lack of technology within the organization.

Case study 3 – working from home

In this case, a project manager in a large consulting firm had a special arrangement to work from home 2 days a week.

This user had one laptop that they used to use in the office for their day-to-day work. However, she never used to carry her laptop back home, and she would use her personal device at home to continue any project-related task.

Every Wednesday, she used to forward pending task emails with project-related documents from her office email ID to her personal email address so that she could work from home using her personal device.

She decided to share some of the project-related Excel files with one of her friends when seeking some help in resolving conditional formatting and calculation issues. While this seemed to be a harmless task, it resulted in the exposure of all the sensitive project-related data on the internet, including data such as accounting-related code, project costs, and customer and employee ID data.

Solution and outcome

Similar to the outcome of case study 1, the organization invested in implementing a zero trust architecture framework along with a **Bring Your Own Device (BYOD)** policy and **User and Entity Behavior Analytics (UEBA)** technology.

Case study 4 – AT&T

The worldwide smartphone revolution has also led to various models of bringing smartphones, data/calling capabilities, and customers together. In some markets, such as India, a customer can buy a phone from any of the tens of brands, buy a subscription from any of the telecom companies, and then use both together. However, in the US and many western markets, the customer enters into a contract with the telecom companies that provide them with a smartphone bundled with data/calling services for a monthly subscription fee, which is a fraction of the cost of the smartphone. The customer benefits by not paying a whole lot of money for the smartphone upfront, while the telecom company gets back its money after a certain number of months (12, 24, or 36). For the duration of the contract, the smartphone is **locked** with the telecom company primarily through technological control exercised by the company. This context is important for you to understand the real insider risk story that unfolded at the company AT&T.

In 2019, the US **Department of Justice (DOJ)** indicted Muhammad Fahd, a Pakistani national living in Hong Kong, and extradited him to the US for "*recruiting and paying AT&T insiders to use their computer credentials and access to disable AT&T's proprietary locking software that prevented ineligible phones from being removed from AT&T's network.*"

Apparently, Fahd had approached a few AT&T employees and offered them a substantial bribe. In return for that, the AT&T employees who agreed to participate in this scam were given "*instructions on how to go about the task, including some lists of smartphone international mobile equipment identity (IMEI) numbers,*" which were then fraudulently and without authorization unlocked. These smartphones could now be used across telecom operators, and AT&T stood to lose millions of dollars in terms of unrealized contractual payments.

The modus operandi to do this allegedly included the installation of some malware on AT&T's internal computers, which helped the conspirators gather information about AT&T's internal information architecture and IT setup. They then used that information to create additional malware that could now process fraudulent unlock requests from remote servers controlled by Fahd and his co-conspirators. This setup was alleged to have been established in 2013–2014, and the scam ran for 4 years till 2017.

There was an additional aspect that came to light in the DOJ indictment. The initial set of employees who were bribed by Fahd and his team were also used to farming more connections within AT&T and recruiting other employees too. The actual number of employees who could be lured is unknown but is certainly more than the few known who were identified in the official investigation.

Solution and outcomes

This is a classic insider threat scenario where a few rotten apples can spread the rot across a company. For a large company such as AT&T, it can be said with certainty that they had protective measures in place, but they turned out to be inadequate. It is important for modern enterprises to monitor employee attitudes and behavior in a non-obtrusive manner. Without taking away from privacy, it is possible to keep a tab on general feelings and intervene only when there is a behavioral spike that demands interest. These behavioral patterns can be further linked to device policy violations and activities such as installing malicious software on company devices to give a complete picture.

The cost of insider threats

The Ponemon Institute's *Cost of Cyber Crime Study* of 2015 contains some interesting findings. Attacks by malicious insiders were the costliest type of attacks and took longer to resolve than other attacks. In the relationship between security budgets and the cost of attacks, things seem to be a little underdeveloped. Even though malicious insiders are the costliest and cause the longest downtimes, only 13 percent of security budgets are being spent on the human layer (reference: *Lord, N.; "Findings From the 2015 Ponemon Institute Cost of Cybercrime Study: The Threats vs. Defenses Gap," Digital Guardian, 18 November 2015 –* `https://digitalguardian.com/blog/findings-2015-ponemon-institute-cost-cybercrime-study-threats-vs-defenses-gap`).

The compounding cost of an insider risk is hard to calculate using any risk management framework. There are several impacts that need to be considered at a high level, such as the value of data, the type of data loss, the equipment damaged, the loss of revenue, the cost of forensic services, the cost of vulnerability mitigation, the cost of compliance and privacy regulation, legal fees, reputational damage, customer notification time, and the money spent.

Summary

In this chapter, we first recapped what we had learned about insider risk and then looked at some real-life examples. Through the real stories from various companies and institutions such as Uber, AT&T, the US government, and an anonymous life sciences company, we could decipher the importance of tracking risky insiders. Across these examples, the core theme seems to be that the companies and institutions were not fully prepared for a scenario where an internal employee could create a huge risk. We also saw some possible solutions and mitigation strategies to reduce insider risk.

All the preceding case studies can be mapped with the MITRE ATT&CK framework TTPs. This framework, along with modern **Security Information and Event Management (SIEM)**, **User and Entity Behavior Analytics (UEBA)**, and **Security Orchestration and Automation Response (SOAR)**, creates a very powerful combination for identifying, predicting, preventing, and stopping all types of insider attacks.

In the next chapter, we will now look at the future of digital risks through the lens of cyber-warfare.

10
Cyberwarfare

War simply means *a period or state of fighting between two countries or entities*. War often means two countries are in an armed conflict. At times, we also talk about war in the context of poverty. So, in popular consciousness, *war* refers to a state where a country, organization, or other entity expends an enormous effort in terms of arms, money, and resources to attempt to be victorious over another party with an opposing point of view.

While war itself is a state, warfare is the engagement model or tactics deployed to win. Cyberwarfare is a new form of this, where computing devices and technology are used to digitally attack an enemy country, organization, or entity. The objective of a cyberwar is to inflict damage on people or objects not limited to military assets in the real world.

Much of the time, cyberwarfare as a term has been used by the media at large to describe how countries such as the United States, United Kingdom, China, Israel, Russia, Iran, North Korea, and others use technology to impact real-world objects, including power and transport infrastructure, or to steal critical data from the military, critical infrastructure, and technology companies.

In this chapter, let us look at the important, grave threat this poses to countries, organizations, and entities in our digital-first world. Let us look at various actors, from cybercriminals to nation-states, and consider the roles played by technology, software, and semiconductors in cyberwarfare.

We will cover the following topics in this chapter:

- Is everything fair in love and war?
- War and its actors
- Advanced persistent threats and examples of such attacks
- The impact of cyberwarfare

Is everything fair in love and war?

Cyberwar refers to a state when a country, organization, or entity (including individuals) executes a digital attack using computing resources to harm another party in either the digital or physical world. The motives for digital attacks can be anything from proving a point to causing interruption to services in the digital or physical world, or disabling the other party's ability to respond to your actions in the digital or physical world.

During the previous two world wars, technology was concentrated in the hands of military and defense organizations. With the rise of the internet since the 1990s, growing innovation in the commercial technology market has mostly become centered around Silicon Valley in the US. New start-ups get rich and grow into massive companies, due to various innovations in software, hardware, the cloud, quantum computing, **artificial intelligence (AI)**, and robotics.

It is interesting to observe that NVIDIA, a leading chip manufacturer in the video games and graphics market, has innovated to create processors that are eight times faster than the processor used in one of the best fighter jets in the world, the F-35. The F-35 had one of the fastest processors available, which earned it the nickname *the flying supercomputer*, but is it not surprising that a video game business division of NVIDIA managed to create a processor that is even faster?

Lockheed Martin is a major American aerospace, defense, security, and advanced technologies company. It's one of the largest defense contractors in the world and provides a wide range of products and services to the United States government and other countries, including military aircraft. Most modern military aircraft use a variety of processors to handle different tasks, such as flight control, navigation, and sensor data processing. These processors are usually highly advanced and powerful, with high processing speeds, low power consumption, and high reliability. You can read more about this by visiting the following pages:

- `https://www.lockheedmartin.com/en-us/capabilities/21st-century-security-networked-solutions.html`
- `https://www.lockheedmartin.com/en-us/products/f-35.html`

The following two photographs show the difference between the processors and motherboards, in terms of size, shape, and capability, used in fighter jets such as the F-35 versus modern consumer GPUs, capable of equal or more computations from NVIDIA:

Figure 10.1 – The F-35 fighter jet processor

The preceding photo can be found here: `https://www.aviationtoday.com/2018/09/27/harris-lm-icp/`.

The F-35's onboard computers can perform 400 billion operations per second. Compare this with NVIDIA's latest processor, the *DRIVE AGX Pegasus*, which can conduct 320 trillion operations per second and is used in electric trucks and cars. That is 800 times more processing power than the most modern fighter jet. Additionally, F-35s will probably feature advanced avionics and sensor systems, which will require specialized processors to handle the large amount of data generated by these systems.

Figure 10.2 – The NVIDIA DRIVE AGX Pegasus processor

The preceding photo can be found here: `https://anyconnect.com/recommended-sbcs/Nvidia/DRIVE-AGX-Pegasus`.

What if NVIDIA technology for games and electric vehicles could be used in a real wartime situation? What if NVIDIA's advanced technology were to start powering cyberwar instead of the gaming industry? I am confident that whenever we have a cyberwar in the future, private companies in the US will be at the forefront. In the ongoing Russia-Ukraine conflict, cyberwar is a prominent area of the battlefield. Today, large US companies such as Microsoft and others have a stronger ability to detect digital attack patterns than most government cybersecurity units. You can get an idea about this from the following headline:

Figure 10.3 – Cybersecurity firms such as Microsoft playing a role in protecting countries from cyberattacks

The preceding page can be found here: `https://www.cnbc.com/2022/02/28/microsoft-says-it-informed-the-ukrainian-government-about-cyberattacks.html`. Thousands of satellites and the sensors attached to them observe objects on Earth in minute detail. What do they show? Today's satellites, as portrayed in Hollywood movies, produce high-fidelity videos. Satellite feeds can use AI-based recognition algorithms to identify humans, battle tanks, missile transportation, and troop movements in real time. There has never been more real-time surveillance done on the planet than now in the name of security. But what if satellites started playing an active role in cyberwar instead of weather forecasting? With the invention and advancement of quantum computing, what if this technology starts making its way into offensive and defensive cyberwar tools and technologies?

Quantum computing

Modern computing is based on calculations using fundamental units – 0 and 1. Most commonly, we refer to these as the state unit of a bit. Quantum computing adds another state in addition to 0 and 1, called a superposition state, that can store both 0 and 1. The bit used to store them is called a qubit. As these qubits can store more, quantum computing provides exponential computing advances in contrast to classical binary computers. So, quantum computers can solve problems that are beyond the reach of even the best supercomputers. If modern supercomputers could solve a given mathematical problem in millions of years, quantum computers would be able to solve those problems in minutes. What if quantum computing were used in cyberwarfare?

A good question to ask is, should governments and private defense organizations accelerate the adoption of technology into cyberweapons? Should conventional weapons such as missiles and tanks be upgraded with autonomous capabilities? In other words, is everything fair in love and war? I cannot predict what will happen in the future, but today I don't see governments and private defense organizations shying away from or slowing down the use of the latest technology to improve conventional weapons. Most private defense players are doubling down on investments in cyberweapons.

What about regulation or control? In the context of nuclear weapons, we have the **Non-Proliferation Treaty (NPT)** (`https://www.un.org/disarmament/wmd/nuclear/npt/`). The NPT helps to slow down and control investments in nuclear technology, along with its destructive use and its transfer to non-nuclear states. Although some shared declarations and initiatives were made at the G7 in 2016 (`http://www.g7.utoronto.ca/summit/2016shima/cyber.html`) with reference to the shared use of cyberspace and the shared commitment to enhance defenses against the cybersecurity threat, we do not see the intent of governments to create a cyberweapons proliferation treaty similar to the NPT that could prevent or ban the use of advanced technology, such as quantum computing, in cyberwar tools.

War and its actors

Throughout human history, there have been numerous wars fought for different reasons, including control of territory, resources, power, and ideology. In the early years, the primary weapons were muskets, pikes, swords, and cannons, which were relatively slow-firing and less accurate. In contrast, the later years saw the development of advanced weapons, such as machine guns, tanks, aircraft, and nuclear weapons, which were far more destructive and had a significant impact on the outcome of wars.

In recent years, warfare has changed due to factors such as media and social media, resulting in proxy wars that involve more actors and the use of digital technologies. Wars in the early years were fought by trained soldiers using weapons such as swords and bows, while in later years, wars were fought with sophisticated weapons and targeted critical infrastructure. The use of bombs and missiles resulted in civilian casualties.

Recent years have brought another dimension of war, cyberwar, which impacts both the military and civilians. Cyberweapons have the potential to interfere with enemy military signals and instructions given to conventional weapons, causing harm to civilian populations through disinformation and denial-of-service attacks on civilian infrastructure.

The emergence of cyberwar in the 21st century brings yet another dimension that impacts both the military and civilians. Cyberweapons have the potential to jam, change, or delete enemy military signals or instructions given to conventional weapons, such as missiles and bombs. Cyberweapons could potentially change the course of missiles or even halt their launch by creating glitches in the missiles' technical components. The impact of cyberweapons on civilian populations includes disinformation such as fake videos, and denial-of-service attacks on civilian infrastructure such as power grids, internet service providers, and consumer logistics systems, resulting in damage to the morale of enemy leadership, military, and the civilian population at large.

Let's see the changes in war actors and their impacts across different eras:

- *Pre-20th-century wars*: People fought these wars using conventional weapons such as swords, bows, and muskets, primarily on battlefields. The primary impact was on the military and not on civilians. An example of such a war is the Battle of Kadesh (1274 BCE).

- *20th-century wars*: These wars were fought with the use of advanced weapons, including tanks, guns, aircraft, and submarines, primarily on land, in the air, and on water. The impact of these wars was not only on the military but also on critical infrastructure. A prime example of this type of war is World War II.

- *Modern 21st-century wars*: These wars are fought using a combination of advanced weapons, cyberweapons, and digital technologies, impacting not only the military but also civilians and critical infrastructure. The battlegrounds are expanded to include land, air, water, space (satellites), and cyberspace across all devices and nations. An example of this type of war is the Russia-Ukraine hybrid war that began in 2022.

It's important to note that the distinction between the different types of wars is not absolute, and there can be overlapping and a mixture of elements from each era in a single conflict, but what has changed in recent times is targeting humans and the use of cyber tools to inflict maximum damage to adversary infrastructure.

More and more countries are developing cyberwarfare capabilities to deal with situations similar to the Russia-Ukraine war. A recent **International Institute for Strategic Studies (IISS)** research paper (available at `https://www.iiss.org/blogs/research-paper/2021/06/cyber-capabilities-national-power`) ranks 15 countries in three tiers in terms of the advancement of their cyber capacities and capabilities:

- **Tier One**: United States

- **Tier Two**: Australia, Canada, China, France, Israel, Russia, and the UK

- **Tier Three**: India, Indonesia, Iran, Japan, Malaysia, North Korea, and Vietnam

Taking into account the preceding research and trying to foresee the future, we can see that cyberweapons will evolve and become more autonomous. In the future, I think humans with authority will stand behind the screen, just in case they want to pause a cyberweapon otherwise acting autonomously. AI will drive war tactics unknown to humans. A mother AI machine or algorithm will control all other machines, with the single goal of winning the war at all costs.

So, are we saying that machines will act in wars based on their own decision-making ability? Machines will decide on military strategies using their experience and conscience. These military strategies will be followed by other machines, cyberweapons, and humans operating conventional weapons, such as missiles, aircraft, and other military hardware. So, the question arises – do machines, software, AI, or autonomous systems have consciousness?

Let us first understand what we mean by autonomy in the context of wars. Autonomy in the armed forces lies with higher-ranking officers responsible for making the right decisions. These decisions are made in the context of the latest information, informed by decision-makers' experience from earlier wars, and directly impact the outcome of a war. The ability to make the right strategic and tactical decisions for the troops is important and impacts the lives of humans engaged in war.

Imagine a war scenario in the future where all human troops must listen to a machine and do exactly what the machine tells them. The machine will become the primary actor and director of war with the authority to make autonomous decisions. Machines will decide which weapons to fire first, and troops will have to obey the machine's decision. Intelligent machines will play the role of the primary actor in upcoming wars, where cyberwarfare will play a critical role in the overall war strategy.

Will humans take orders from machines?

Forget the context of war – let us just see whether humans are receptive to taking orders from machines in general. Let us say I am driving my car and the machine in my car judges my driving. Modern cars give us voice warnings when we drive without wearing a seat belt. Intelligent car systems warn us when we are driving beyond the speed limit. Personally, I do not like cars commenting on my driving skills, and warnings such as speed-limit notifications can be turned off. The question is, should I trust, listen to, and follow car system instructions?

Guess what? We, like obedient students, follow what software tells us when it is Google or Bing Maps. We also often do not realize when we use ride-hailing apps such as Uber, Lyft, or Grab that the actor we obey the instructions of and make payments to is a piece of software. Yes, Google Maps is software, a kind of machine in the cloud that sends graphical messages with maps to your screen. Yes, read it again – it is a machine. Ride-hailing apps are also software or machines. Put simply, Uber is a machine that does not have a physical body. You interact with this machine using your phone, and what you see is an app that you click. Once the app starts, it has a map and a set of functionalities for you to operate. The actual ride-hailing machine is an array of computers – thousands of servers with fiber cables and lots of software in a data center maintained by humans and robots. Ultimately, it is an AI that finds a car and a driver, orders them to pick you up from your stop and drop you at your destination, and asks for money from you as payment. It then takes its commission and pays the rest to the driver. Both the driver and the passenger *obey* the ride-hailing app. So, the actors in the ride-hailing ecosystem changed from the *rent-a-car* or taxi business that operated on human-to-human interaction. In the Uber world, a human is just an actor, and which car you choose, how much you pay for your journey, and your route are decided by a new actor, the AI software, and *that* is not a human.

The AI behind a ride-hailing app tirelessly matches available taxis, cars, and drivers with potential riders. Machines can perform tasks that are too dangerous or difficult for humans to do, such as working in hazardous environments, driving in difficult terrain, or performing repetitive tasks with high precision. They also do not get tired or require breaks, so they can work continuously for long periods of time. Additionally, machines can be programmed to work around the clock, which can increase productivity and efficiency. They can also be customized to perform specific tasks, such as airport pickups and drop-offs for passengers. Finally, they don't need to be paid or have benefits such as health insurance, vacation time, and pensions. It is more efficient to develop a capable machine as compared to hiring several smart people to do the same job.

Similarly, in the case of a cyberwar, machines will be capable of doing a lot more than hired human soldiers and have a higher degree of execution ability (automation). Let us take this to the next level. Let us give machines an algorithm so that they can think and make decisions on their own – for example, making battle tanks intelligent and autonomous. I will let the autonomous tank assess the situation and make decisions on its own. This changes the battlefield scenario completely, as machines will play independent roles. Humans do not have to sit inside an autonomous battle tank to operate it, aim a gun, or fire ammunition. Autonomous battle tanks will do what they need to destroy the enemy.

Based on all of this, I conclude that humans will indeed listen to machines in war, just as today we take alternative routes when advised by Google Maps.

Advanced persistent threats

In the previous section, we learned the role of machines as new actors in war. We also learned how cyberwarfare is becoming critical and influencing overall war strategies. In this section, we will build on what happens when machines become smart and intelligent.

When I say machines will play a leading role in future wars, most of us will visualize those machines as some form of hardware. Let me break that model of thinking. A machine can be either hardware or software. As I described in the previous section, a ride-hailing application on your phone is a machine in the form of a software mobile application.

Now, if machines can morph into software, it is easy for them to travel via computer networks. These machines can move across organizations, countries, and geographic boundaries in split seconds, just like a computer virus. They can move faster than supersonic missiles. These machines are also more intelligent so that they can take multiple actions based on the environment in which they are propagating. This makes these machines lethal. They can act autonomously and move across geographical boundaries in seconds. The threat this presents to the enemy is persistent, as the software can take autonomous decisions to reach its goal of destroying or infecting enemy infrastructure.

The computer industry calls such machine threats and cyberweapons **advanced persistent threats (APTs)**. APTs in the form of software are called malware.

Catching the owner of an APT can be difficult for a number of reasons. First, APTs are often created and distributed anonymously using the internet, making it hard to trace the source. Second, APT creators often use sophisticated techniques to hide their tracks, such as using **virtual private networks (VPNs)** or the Tor network to conceal their IP addresses. Additionally, APT creators can also use multiple layers of encryption to protect their identity and location. Finally, many APT creators operate from countries with weak cybercrime laws or inadequate law enforcement resources, which can make it difficult for authorities to bring them to justice.

If an APT creator is caught, they almost always deny their involvement in the creation or distribution of the APT. This can make it difficult for law enforcement to prove their guilt in a court of law. Some APT creators may also claim that their APT was created for good purposes, such as for research or testing security systems. In these cases, it can be difficult for authorities to prove intent to harm. Additionally, some APT creators may also claim that they were not aware that their creation was being used for illegal activities and that they did not intend for it to be used for harmful purposes. Let us select a few of them and study the role played by machines as software actors and the impact these APTs have had in the real world.

The Colonial Pipeline attack

Let us go back to 2021. Colonial Pipeline is a private company in the US responsible for supplying oil across eastern states from Texas, Georgia, and Alabama right up to Pennsylvania. No one predicted that a software-based cyberattack would penetrate all of the physical security measures in place to protect the organization.

Cyberattackers created malware that encrypted the company billing software, halting the ability to generate a bill or an invoice. As this happened, most gas stations, even though they had enough gas supply, were unable to deliver gas to customers. Gas stations expecting stock refills were not able to place orders and ran out. Customers, on other hand, noticed this blip in supply and sought to stock up on gasoline, resulting in long queues and panic buying, which further impacted the delivery of gasoline to military and aviation customers.

This APT was unique, as it had a direct impact on citizens and consumers of petroleum. This was a class of APT that encrypted a corporate system vital to the supply of gas to customers and gas stations. This attack showed how you can digitally disrupt multiple gas stations and halt their operations with the press of a button.

The Shamoon virus

In 2012, Iran's Shamoon virus wiped 30,000 hard drives at Saudi Aramco. A quick search online reveals that a group of hackers calling themselves *Cutting Sword of Justice* claimed responsibility for the incident. The reason they launched this attack was to retaliate against the Al-Saud regime for what the group called widespread crimes against humanity. Shall we call it a digital protest?

In this class of APT or malware, the idea is to demonstrate control over enemy infrastructure and delete files, broadcasting a public message as a digital protest and making it difficult for the enemy to respond.

Stuxnet

Let us jump back to the year 2010. Operating under the codename Operation Olympic Games was what we in the cyber world now know as Stuxnet. Stuxnet used an advanced persistent malware called **Flame** to disable Iran's nuclear program by penetrating and inserting malware in its computer networks. The Stuxnet family of APTs also have other payloads, such as Duqo, Havex, Industroyer, and Triton, that have been used over the years to target other countries and industries.

Stuxnet attacked Iran's nuclear program, which operated centrifuges to enrich uranium. The Flame payload in the malware hacked into the centrifuges and made them spin at far greater speeds than they could physically withstand. It is believed to have been created by a nation-state with the intention of disrupting specific **industrial control systems** (**ICSs**), specifically those used in nuclear facilities in Iran. It specifically targeted Siemens **Supervisory Control and Data Acquisition** (**SCADA**) systems, which are used to control and monitor industrial processes in a wide range of industries.

The worm was able to spread via removable media such as USB drives and between computers on a network. Once it infected a system, it was able to identify and target specific systems, such as those used to control the speed of centrifuges at the Natanz nuclear facility in Iran, causing them to spin out of control and self-destruct.

The impact of Stuxnet was significant. It was able to disrupt the operations of the Natanz nuclear facility, causing significant damage to the centrifuges and setting back Iran's nuclear program for several years. It also served as a wake-up call to the vulnerability of ICSs and the potential for cyberattacks to cause physical damage and disruption. It also highlighted the potential for nation-state cyberattacks and the use of malware as a weapon.

I would say this was a very lethal idea and a well-executed attack. It involved computer APT malware creating environmental conditions electronically that resulted in the closure of a uranium enrichment facility, an essential part of Iran's nuclear program. It is like winning a war without firing a shot.

Operation Desert Storm

Operation Desert Storm was the codename for the military operation led by the United States and a coalition of other countries to expel Iraqi forces from Kuwait in the early 1990s. The operation was in response to Iraq's invasion of Kuwait on August 2, 1990, and was authorized by the United Nations. The coalition, which included countries such as the United Kingdom, France, and Saudi Arabia, began a massive air campaign on January 17, 1991, followed by a ground invasion on February 24, 1991. The ground invasion was quick and successful, and Kuwait was liberated within 100 hours.

After the invasion, Saddam' Hussains forces ran a captive fiber network that the coalition forces bombed as well as deployed and positioned a satellite overhead. Saddam switched to his backup system using microwaves, which the US monitored and even eavesdropped on to disrupt and misconfigure the system. Coalition forces used electronic warfare to disrupt the communication and command and control systems of the Iraqi military, making it harder for them to coordinate their operations, which included passing them erroneous messages.

This helped coalition forces disrupt Iraq's radar and counter its military by becoming informed of potential moves and actions the Iraqi adversary was planning to execute. Desert Storm introduced cyberwarfare tactics to disrupt the Iraqi military's command and control systems, communications, and logistics.

In this class of APT or malware, the idea is to create interference so that the enemy is forced to switch to a backup system for which monitoring was already set up.

Impact of cyberwarfare

Today, most of us think of a cyberattack as when a website, database, or server is attacked or taken control of by cybercriminals. That is not the whole story, as seen in the previous examples of cyberattacks I shared, which showed how their impact can range from restricting a nuclear program to affecting a company's ability to sell oil to consumers at gas stations.

A camera in an elevator, a smart TV in a hotel lobby, and some temperature-monitoring equipment at the entrance to a building all are connected to the internet. They are all susceptible to cyberattacks *and* capable of participating in cyberattacks.

Contemporary cyberwarfare is the use of any and every form of digital attack on your enemy. Gartner predicts that by 2025, cyberattacks will have weaponized operational technology to harm or kill humans (read about the study here: `https://www.gartner.com/en/newsroom/press-releases/2021-07-21-gartner-predicts-by-2025-cyber-attackers-will-have-we`). As we saw in the examples provided, the cyberattack on Colonial Pipeline resulted in gas supplies being disrupted and power outages leading to the disruption of business and wider human life. Cyberattacks on critical health infrastructure not only affect individuals' data but also wreak havoc in the physical world. All these disruptions in the physical world indicate a change in the demography affected by war. Cyberwarfare is about creating a direct impact on the lives of citizens at large, causing panic and stress in the wider population. This will lead to attacks on both military and civilian infrastructure, stretching enemy resources and attention. Most of this disruption will inflate prices and create shortages of essential goods, inflicting financial stress upon the enemy. Even worse than that, these cyberattacks may soon turn deadly. The cyber battlefield has expanded, and the stakes are higher than ever before.

The message is clear – when countries fund or sponsor such cyberattacks, it's a part of cyberwarfare. Rapid digitization has created this new warfare, bringing about these invisible weapons in addition to all the productivity software and services that we are all familiar with.

Summary

In this chapter, we traveled back to earlier times when wars were fought with swords, knives, and daggers. The 20th century brought industrialization and factories. These factories manufactured weapons such as guns, tanks, and missiles. Guns were not new, but advancements in technology created rapid-firing machine guns. Tanks as armored combat vehicles equipped with cannons and machine guns were more effective than rifles. Air superiority then became critical to victory. Aircraft with bombing capabilities changed the nature of war during this era. Now, what is changing in the 21st century is the addition of cyberwarfare.

This cocktail of kinetic war and cyberwarfare is lethal, as it impacts the lives and infrastructure of civilians. In the future, dependence on digitization for payments, communication, food supply, and transportation will be exploited, causing disruption to social life and business. As much as I enjoy technology, the dark side of autonomous weapons and the weaponization of cyberwarfare impacting civilians scare me deeply.

In the next chapter, we will examine details of governmental and institutional regulatory risks. We will discuss how regulatory risk can impact digital transformation across various industries, along with some case studies and frameworks to better understand contemporary regulatory risks.

An Introduction to Regulatory Risks

To understand the landscape of regulatory risks better, we have created a fictionalized story based on some real scenarios. The story begins with an email from the legal advisor to the CEO of an e-commerce company that is seeing rapid growth post-pandemic. This is what the email looks like:

Dear XYZ,

Congratulations on the fabulous Q1 numbers and thanks a lot for your stewardship, especially in the current challenging times post-pandemic.

I am writing this email in continuation with my previous couple of emails regarding the latest National Consumer Regulatory Body circular. The body advises us to:

- *Define all customer information together with exchanges and transactions with customers*
- *Define parameters to preserve this information until explicitly warranted*
- *Identify all employees who deal with confidential and customer data*
- *Execute a program to safeguard all of this information and protect customers*
- *Put in place audits for the same*
- *Have a system to produce information when asked for by the regulator*

Thanks to your leadership and the consistent efforts of the sales and marketing teams, we are seeing an unforeseen number of new customers and our transaction volumes are also exploding. But we are very constrained in terms of resources to follow regulatory advisory. This advisory is a precursor to the coming mandatory regulation, and I am worried that if we delay further, we shall run into high waters within a few months as soon as the regulations are passed.

The CEO, XYZ, took a deep breath and pinged her deputy, her COO. As the leader of an e-commerce firm, she felt like she had come out of hell quarter after quarter and was barely keeping afloat. Theirs was a D2C consumer products company; they had three venture-backed smaller start-ups breathing down their necks while they were still locked in a battle of attrition with large traditional manufacturers, who were going all the way to retain their large base of customers and trying to fight for the new customers with much larger paychecks. While reading this email, she realized that she had ignored the two other emails that her legal advisor had sent. It is not as if she had not thought about this earlier, but her thoughts always veered on three aspects – the regulations are still not binding, the competition is already outspending them on new customer acquisition and sales team hiring, and they are getting close to an IPO, so she needs to ensure that they grow at least 2x before they announce an IPO so that shareholders can get their due. In this scenario, where each dollar was precious to meet business goals, she did not see the urgency to spend on resources and tools for something that was not legally binding yet. She knew her employees and she trusted them to do the right thing as they had amply proved in the past.

The point of this story is for you to be able to empathize with situation leaders who frequently find themselves in such current scenarios. As you read this chapter, you will get more context on where such scenarios arise and possibly think about regulations and compliances in the same breath as business metrics and outcomes. With that anonymized story in place to give you a practical understanding, let us look at the theoretical framework of regulatory risks.

Regulatory risks – an introduction

Most business leaders will empathize with XYZ. Business leaders often find themselves at the forefront of dealing with competitors, customers, and partners and struggling to conjure enough resources to deal with all of them effectively.

Typically, compliance and regulatory risks have been associated with certain kinds of businesses, such as those dealing in sectors such as BFSI, manufacturing, or the public sector. Companies that deal with international operations or operate in highly regulated sectors have highly developed practices to assess and mitigate risks that emerge out of the regulatory landscape, both on a short-term and long-term basis. Take, for example, any multinational corporation that frequently recognizes revenue across various countries around the world. The currency exchange rates can vary dramatically on a day-to-day basis, subject to various regulatory, political, and economic signals. Most such organizations look at regulatory risk from a cash management viewpoint and hedge their bets via various financial instruments. As another example, you can look at BFSI sector companies. Almost all of them are governed by a clearly defined set of guidelines by a regulatory body in the geography they operate in. Many of their practices emerge from the regulatory definitions and many of their functions need to converse with the set of regulations prevalent. The following figure provides a very basic view of the regulatory risks a bank operates under (please note that this is not exhaustive and purely illustrative):

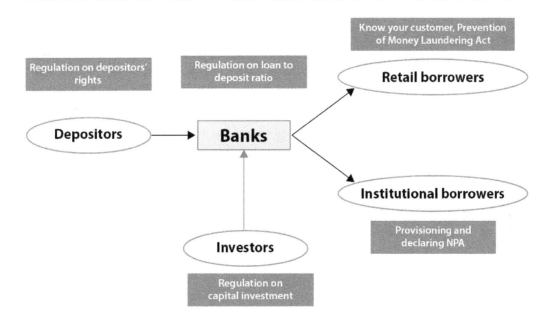

Figure 11.1 – A simplistic view of regulations for a bank

The red boxes in the preceding figure mention one among the many regulations that a bank is governed by for each of its stakeholder relations and transactions. Having given you an overly simplistic view of the regulations for one specific industry, let us look at how the regulatory landscape is shifting post-pandemic.

Digitization and the expansion of the regulatory risk landscape across sectors

This book talked about the rapid digitization during and after the pandemic in detail in the previous chapters. In this pre-pandemic era, you could excuse a business leader who thought about regulatory risks once a year just before their board meetings, but at the time of writing, irrespective of the sector or the industry a company operates in, all leaders need to be aware of the regulatory risks. Not only that, but it is also contingent on them to ensure that a culture of compliance percolates through their organization and that they support that through the essential technological tools as well.

The world of business has changed quite a bit with the advent of digitization. With the pandemic acting as a catalyst, the whole rate of change has accelerated exponentially. This acceleration has led to the emergence of new business models, which has, in turn, made the regulators sit up and take notice. At their core, regulations are a means through which the society, governments, and socioeconomics of a region communicate their expectations to the stakeholders. The current phase of digitization is disruptive for regulators from three dimensions – speed, scale, and interdependence.

Speed

Digitization has reduced the time it used to take for a business to go from zero to one both in terms of employees and revenues. It has also reduced the time taken to serve each customer, be it through app-based services, hyperlocal deliveries, or ridesharing. Regulators have the unenviable task of ensuring that customers, employees, and laws are protected from being a casualty to this speed.

For example, here is a view of top tech companies of the world and how each company has superseded the other in terms of growth to a billion users (you can find this graph here: `https://bit.ly/VisualCapitalistLink/`):

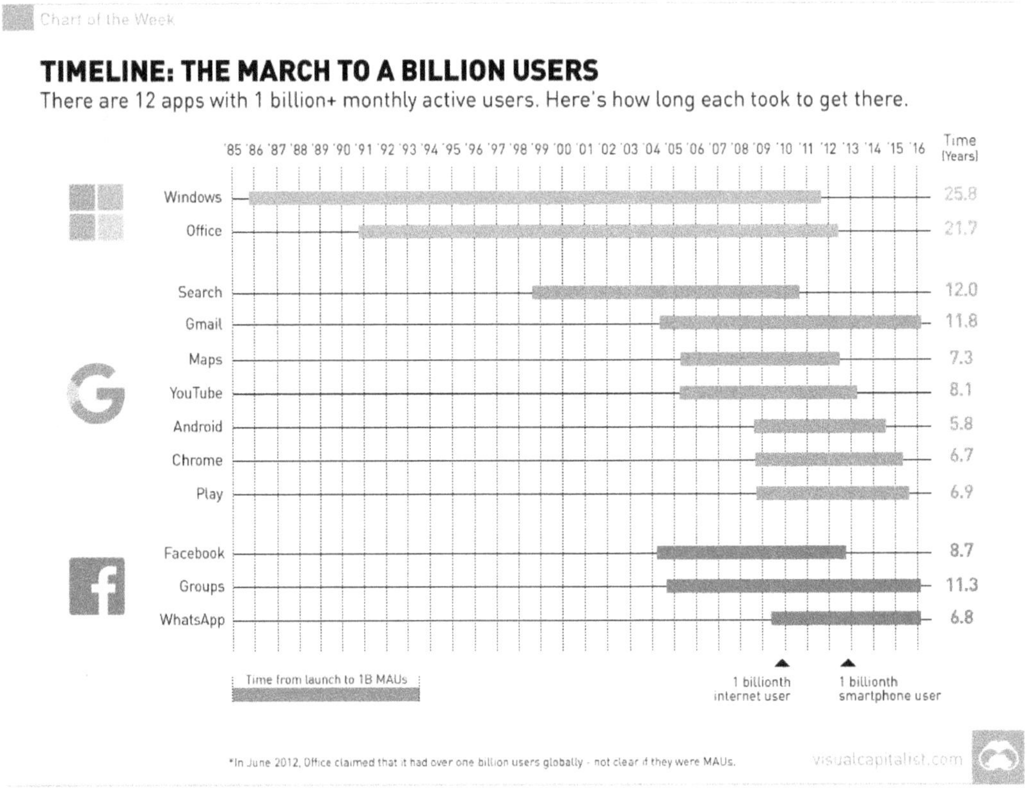

Figure 11.2 – Time from launch to 1 billion monthly active users

Here is another way of looking at speed. Let us just look at the recent unicorns and the time that they took from the day they were founded to the day they touched $1 billion revenue (source: `https://bit.ly/BusinessInsiderReport`):

Fast starters: start-ups with fastest route to $1 billion revenue	
COMPANY	Years from founding to $1 billion
Slack	1.25
Groupon	1.46
Akamai Technologies	1.58
Xiaomi Technology	1.71
Calient Technologies	2.02
Yello Mobile	2.28
Twitter	2.32
Hangzhou Kuaidi Technology	2.43
Pinterest	2.46
Webvan Group	2.56
Instacart	2.58
Lazada	2.75
Inder	2.83
Zynga	2.88
Square	2.93
Better Place	3.07
WeWork Companies	3.09
Snapchat	3.09
Uber Technologies	3.11
Hortonworks	3.23

Table 11.1 – Start-ups that were the first to reach 1 billion dollars in revenue

Now, let us look at the scale dimension of the digitization impact on the landscape.

Scale

With fast speed comes rapid scaling. There are hundreds of companies in India, for example, that serve more than 10 million customers daily. Rapid digitization leads to billions of transactions happening across businesses every day. For a regulator, it is almost a nightmare to ensure that all of these transactions are being completed in a manner that agrees with their governance ideals and their citizens' expectations.

For example, if we just look at the top 25 social media and communication companies by usage (source: `https://en.wikipedia.org/wiki/List_of_social_platforms_with_at_least_100_million_active_users`), we will see that all of them are serving more than 100 million users per month. The amount of personal, confidential, and sensitive data that is entrusted by common people on these platforms is humongous. Regulators have the duty of thinking proactively and making sure that the average user is secured while using these platforms and that the companies that make these products take abundant precautions to comply with their regulations:

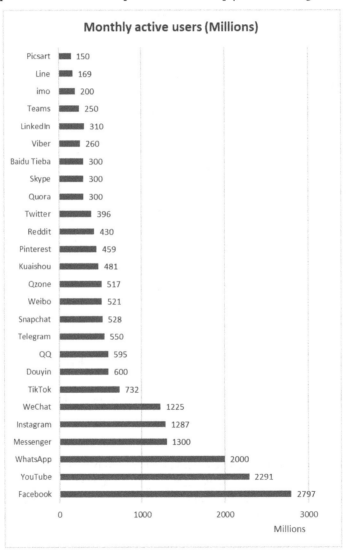

Figure 11.3 – The top 25 companies of the world by monthly active users (data as of 2020 or before; numbers might have changed in the interim)

Having looked at the dimensions of scale and speed, in the next section, we will look at the dimension of interdependence, which shifts the magnitude of the impact of both scale and speed.

Interdependence

To help companies acquire customers with speed and at scale, the digital era has spawned various intermediaries who help companies at various stages in their business. Think of the last item you bought from an e-commerce store, where the vendor hosted their catalog on a marketplace site that you accessed from your browser. Clicking on that item started a whole host of activities. Let us understand this by looking at an example – a logistics management intermediary tracked availability and provided it through an API to the e-commerce site; a SaaS service used by the vendor stored your order and a payment gateway intermediary handled your credit card records before your order was confirmed. The job of a regulator gets even more complex in a scenario like this.

In their research paper titled *The evolution of platform business models: Exploring competitive battles in the world of platforms* (https://www.sciencedirect.com/science/article/pii/S0024630118306368), Zhao et al. delved deeper into the multi-sided platform business models. One diagram in their paper is very helpful for delving into our point of interdependence:

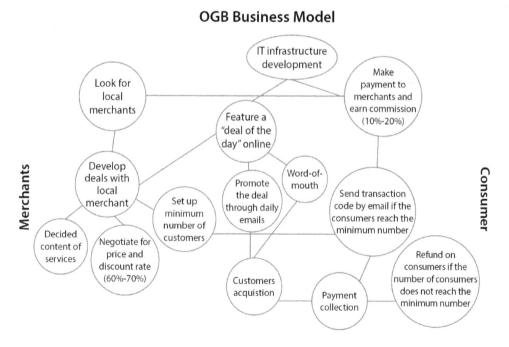

Figure 11.4 – Multi-sided platform business model (example of Groupon)

At the time of writing, countries such as India, South Africa, and Thailand have personal data protection bills in various stages of implementation. In European and American parliaments, there is a lively debate around various implications of digital business models, which will probably result in a whole host of new regulatory measures. And the cost of being unaware or not fully compliant with those might be prohibitive for a lot of companies.

Now, let us move on to the next section, where we will try to create a framework that applies to the majority of modern businesses and explains their interaction with regulations.

A framework to understand modern regulatory risks for all businesses

Any modern business has multiple stakeholders and their compliance with regulations is mandated across those stakeholders. The next diagram captures a small set of such dependencies to paint a picture for you:

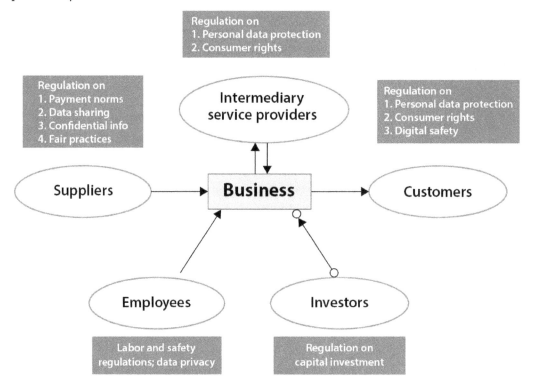

Figure 11.5 – Modern businesses and regulations

As the preceding diagram shows, any modern business has multiple stakeholders, and all those stakeholders and how they operate together come under the ambit of multiple regulations. To understand this further, let us look at a sample of regulatory risk types in the following table:

Regulatory Scope	Stakeholders Impacted	Explanation
Trade Policy	Customers, partners, employees	Nations and governments determine specific geographies or sectors where companies can or cannot operate. For example, trade between warring nations is banned by the regulations in those countries and a company runs the risk of severe sanctions if any of its stakeholders engage in trade with a customer or partner operating in the enemy state.
Corruption	All	All transactions made between the stakeholders are expected to be fair and non-corrupt.
Taxation	All	All transactions between the stakeholders of a company are expected to be compliant with the due taxation laws.
Health and Safety	Employees, customers	Nations and regulators have specified mandates that need to be adhered to, to ensure the health and safety of the employees and customers of a company.
Data Privacy	Employees, customers	More than 70% of the countries around the world have some regulation to ensure that employees and customers have rights over their data.
Environment	All	If a company deals with natural resources or deals with products that impact the ecology around a nation, they are governed by environment-specific regulations.

Table 11.2 – Prominent regulatory topics

Consequent to the broad archetypes of regulatory topics, a company faces many negative outcomes if they don't comply with these regulations.

Let us go back to the CEO of the e-commerce company from the beginning of this chapter. She went on to talk to her COO about the email from her head of risk. The COO shared with her that he has already seen a report from one of his employees, showing that a supplier of theirs has sent a mass mailer to the customers after taking a copy of their order data. While they have been warned and this episode was caught, it might be impossible to catch all such infractions, especially when they are growing 2x every quarter.

Summary

In this chapter, we saw how modern businesses of all shapes and sizes are coming under the ambit of multiple regulations. These regulations cover all their stakeholders and it is pertinent for every business to be compliant in all the ways that regulators mandate them to. In the next chapter, we shall delve deeper into how governance and compliance have emerged as disciplines in response to these regulatory requirements.

The Evolution of Risk and Compliance Management

Risk management and compliance management are well-established disciplines today. However, the practice of these disciplines and even awareness of them is not distributed equally across the world. This chapter aims to provide a common baseline for understanding risk and compliance management for you.

We shall start by defining *risk* and our traditional understanding of it. The evolution of our understanding of risk is important to know as that sets the stage for understanding the contours of *risk management* as a discipline. You will then read about some real events that destabilized the economic system and how regulatory bodies and governments responded to them. You will discover the need for companies to treat regulatory risks specifically, and you will understand how the digital and IT landscape of companies is at the center of all of this.

The content of this chapter will help you understand the following:

- What is risk?
- How was risk traditionally managed?
- A timeline of missteps by organizations
- A timeline of how regulators responded
- The evolution of the current compliance landscape
- Risk management to GRC to compliance management
- What changed with COVID?

What is risk?

Compliance management deals with the management of an enterprise or institution's risks of failing the various regulatory norms and frameworks, which might lead to massive loss of money, prestige, reputation, or overall business. Before we start talking about compliance and the evolution of compliance management as a way to deal with regulatory risks and much beyond, let us focus on the word *risk*. The Merriam-Webster dictionary defines risk as the following:

1. The possibility of loss or injury: *peril*

2. Someone or something that creates or suggests a hazard

3. The chance of loss or the perils to the subject matter of an insurance contract. Also:

    *   The degree of probability of such a loss

    *   A person or thing that is a specified hazard to an insurer

    *   An insurance hazard from a specified cause or source

4. The chance that an investment (such as a stock or commodity) will lose value

For centuries, traders and business folks of all kinds have known that almost all businesses come with a modicum of risk. To a large extent, the evolution of humanity can be seen easily through the evolution of the way we assess and manage risks in our day-to-day lives. And businesses have been the most integral parts of our day-to-day lives since we came together as a society.

In an interesting recent podcast, Joseph Henrich (a professor of human evolutionary biology at Harvard) touched upon the evolution of *big gods* in human societies and how their origins could be traced to arbiters of the transactions between trading communities. Ancient human beings saw natural calamities, geographical distances, and human behavior all as vectors of high risk, and religion was, to a large extent, a risk management tool. They progressed to find better tools to manage many kinds of risks, but fundamentally the reality remains the same – things with a very small chance of happening do happen. And for every kind of mishap that we can factor into our scenario mapping, there are many others that we don't even know exist. This is the struggle between *known unknowns* and *unknown unknowns* that we have just begun to even articulate.

On the topic of known unknowns and unknown unknowns, it is opportune to also mention Nassim Nicholas Taleb, who shot to prominence with his book *The Black Swan – The Impact of the Highly Improbable*, published in 2007. He established the metaphor of a *black swan* – swans are almost universally white, and till the first black swan was spotted somewhere, nobody was aware they even existed. This made the public aware of unexpected events whose potential to happen only becomes clear once they have happened. Not to forget the average white swans, which we are all aware exist and still need to look out for in order to spot. This is the landscape of risks. Former SEC chairman Harvey Pitt once famously said, "*Rare events aren't all that rare. Lightning does strike twice, and the unimaginable occurs more frequently than most of us believe. It's important to expect the unexpected—while that may seem as if it's an oxymoron, it's actually a good prescription for avoiding cautionary tales.*"

If the inevitability of risk is so ubiquitous, the urge for human beings to respond to it is very natural. In the next section, we will look at the origins and evolution of risk management.

Origins and evolution of risk management

In the previous section, we talked about how ancient social and religious customs emerged out of the necessity to manage risks and establish trust. But like every other field of knowledge, risk management got a boost between the 15th and 17th centuries during the Renaissance. The technique of double bookkeeping was invented by a mathematician and monk named Luca Pacioli, which helped businesses manage their accounts, and even the most modern accounting frameworks are based on that. Luca Pacioli was a brilliant mathematician and collaborated with Leonardo Da Vinci on several mathematical frameworks, puzzles, and tricks. But for all his brilliance, it was a problem he could not solve that perhaps provided the biggest impetus to modern risk management. The question he had posed for other mathematicians of his era was related to deciding the outcome of an abandoned game of chance between two players, especially when one player was ahead at the time of abandonment. Without going into the details of the question, it is sufficient to say that for 200-odd years no mathematician could solve this puzzle. And then a French nobleman posed this as a challenge to Pascal in 1654. Blaise Pascal was one of the most famous mathematicians and scientists of that era (even to date, some of his theorems find prominent mention in science and engineering), but he also had to seek help from another famous lawyer and mathematician, Pierre Fermat. And not only did they solve the problem together, but they also came up with the basic laws of probability that enabled a whole new way of looking at risk.

It was around the same time that two other remarkable gentlemen, one of them a hobbyist (John Graunt) and another a scientist (Edmond Halley) did some fascinating work that blended into the theory of probability and strengthened it further. It was around statistical sampling, probability theory, and an estimation of the number of deaths/births in a specific year that their focus lay. Perhaps it won't surprise you that this coincided with the emergence of a formal insurance industry around the same time and went on to become the first risk management system of the modern age.

We have to take a leap in the timescale and look at the Second World War as the seminal point when a lot of practices and nomenclature we use for modern risk management came into place. The discipline mostly began by using insurance to manage risks, but from World War II to the late 60s, various companies and industries saw new risks being created, old risks being escalated to new levels, and in order to deal with them adequately, risk management evolved as a modernized discipline. In the next section, we will see how the broad umbrella of risk management started seeing the emergence of sub-sections that tended specifically to regulatory risks.

From risk to compliance management via increased digitization

Governments and regulators always had an eye on the economy and the businesses operating within that. The two-way communication between businesses and governments/regulators happened through policy-making and common legal enforcement. However, two moments in a 24-year span pronounced the need for government and regulators to take a more defined view of business practices and, in turn, spurred the business community to evolve with cognizance of those regulatory concerns. The first was in the 1970s when the US Securities and Exchange Commission found about 400 US companies indulging in corrupt practices in their business operations outside the US, which included making questionable payments to political agents or government officials. This led to the *Foreign Corrupt Practices Act* of 1977. In response, five major professional accounting associations in the US – the **American Institute of Certified Public Accountants (AICPA)**, the **American Accounting Association (AAA)**, **Financials Executives International (FEI)**, the **Institute of Internal Auditors (IIA)**, and the **Institute of Management Accountants (IMA)**, came together to organize a joint initiative to fight corporate fraud. The body was called the **Committee of Sponsoring Organizations of the Treadway Commission (COSO)**. It came up with the COSO framework, which framed business practices and processes into an audit lens through *internal controls*. The COSO **enterprise risk management (ERM)** model has become a very influential framework despite not being mandated. We would encourage you to check out `https://www.coso.org` for more information. The second moment was the Enron scandal in 2001. Its impact was worldwide and in immediate terms, it led to the *Sarbanes-Oxley Act of 2002*, which was designed to protect investors from fraudulent financial reporting by corporations.

In other geographies such as Europe, APAC, and India, the global financial crises and multiple internal catastrophes have guided the local regulators to attempt to stay ahead of turbulence that can impact companies and investors. The German Stock Corporation Act and India's Companies Act have, for example, both expanded and now include additional responsibilities on companies' boards to set up mechanisms to monitor and report on risks that put their existence in danger. The **Organization of Economic Co-operation and Development (OECD)** is an organization comprised of 38 countries from across Europe and other continents (such as Japan, Mexico, and Singapore). It has been at the forefront of keeping a watch on internal and external trends impacting its economies and has published various guidelines that have helped its corporate governance landscape as well as its risk management practices.

As you will notice, the regulations started from accounting practices but evolved with the digital landscape of companies – later-stage regulations started taking into account the IT and cyber practices of companies that could cause risks to consumers and investors. Digitization has become the pillar around which companies build for growth, and regulators aim for controls. From a compliance and regulatory management perspective, digitization had to be seen from two lenses – data and processes. We have delved deeper into the *data* aspect in *Chapter 13, The Role of Data and Privacy in Risk Management*. Digitization has fundamentally changed many business processes, and regulators have always been cognizant of this. We also touched upon it in *Chapter 11, An Introduction to Regulatory Risks*.

To summarize, the evolution of compliance management from risk management has been a sequential product of the following three things:

1. Major missteps by businesses or sectors leading to scandals or public losses, prompting government and regulators to take action

2. Governments or regulators coming up with regulations to ensure the rule of order, including keeping an eye on the foundational changes that the digitization of businesses was driving

3. The need for companies to evolve their practices to stay on the right side of regulations and keep track of their compliance posture

In the next sections, we will take a look at some of the most famous examples of the first two points in the previous list, then look at the third in terms of practice.

A timeline of the top events that made regulators take notice

We have created a timeline to show some of the major events which made the regulators and governments sit up and take notice. As we mentioned earlier, many regulatory developments have been a consequence of real-world incidents or mishaps on a large scale:

- **Enron (2001)**: Enron invested in a new business line of video on demand. Their endeavor guzzled a lot of money and did not give any returns. The company's management started cooking its books to conceal its losses from this new line of business and started showing yet-to-be-made profits as actuals. In a way, this was a trendsetting example for the 21st century that set the alarm bells ringing for regulators all around the world.

- **Worldcom (2002)**: This was, at the time, the largest corporate scandal in the US. In order to manage stock prices, the company's CEO led a senior team to overstate the earnings from 1999 to 2002. It was caught by a team of internal auditors in 2002 when they found about $3.8 B of fraudulent entries on the balance sheet.

- **Siemens (2004)**: In 2004, there were widespread allegations of bribery involving Siemens AG and Greek government authorities around the purchase of some equipment in the late '90s. The amount involved was supposed to be in the 100's of millions, and it caused massive public consternation in Greece. The scandal concluded after the company settled for more than $1 B in fines to the US **Securities and Exchange Commission (SEC)**, the US **Department of Justice (DOJ)**, and German authorities.

- **AIG (2006)**: In 2006, AIG settled with the SEC by paying $1.6 B to resolve claims related to improper accounting, bid rigging, and practices involving workers' compensation funds.

- **The financial crisis (2008)**: Cheap credit and very relaxed lending standards fueled a housing bubble. Financial institutions traded different varieties of these risky loans as a product, and when the bubble burst, many of them found themselves holding trillions of dollars of mortgages that were worthless. This triggered a worldwide recession with a huge human impact.

- **Walmart (2011):** Walmart failed to maintain internal controls on the **Foreign Corrupt Practices Act (FCPA)** during the expansion of its international operations in Brazil and settled with the US DOJ by paying $137 M as a fine.

- **Yahoo (2014):** In two events in 2013 and 2014, the largest known breach in the history of the internet happened at Yahoo. The data breach impacted almost 3 billion of its user accounts and was not discovered till as late as 2016/17. This resulted in multiple ongoing government investigations against Yahoo and also caused a loss in terms of the final value of its sale to Verizon.

- **Theranos (2015):** Theranos was a Silicon Valley startup that garnered a lot of investors and media attention with claims of having devised blood tests needing a very small amount of blood in a rapid amount of time, thus revolutionizing the testing and health tech industry. After drawing $700 M in funds from **venture capital (VC)** firms and investors and reaching up to $10 B in valuation, they were found to have completely faked their claims.

- **Wells Fargo (2016):** In 2016, Wells Fargo employees created millions of fake savings accounts on behalf of their clients without taking their consent. This fraud became visible after various regulatory bodies investigated and fined the company about $185 M for what they deemed as *illegal activities*. The company faced additional civil and criminal litigations, which could impact it to the tune of $2.7 B.

- **Equifax (2017):** Between May and July of 2017, Equifax, an American credit bureau company, was breached by attackers who were able to exfiltrate personally identifying data of 147 million of its customers. This led to multiple litigations and investigations, and the company agreed to a global settlement with the **Federal Trade Commission (FTC)**, the **Consumer Financial Protection Bureau (CFPB)**, and 50 US states for $425 M.

- **Tesla (2018):** Tesla settled an enforcement action with SEC in 2018, where it was alleged that its chairman, Elon Musk, had committed fraud by tweeting about a prospective buyout. There were a few other social media posts that led the SEC to reprimand the company, and ultimately, Elon Musk had to resign from the chairman's post, along with paying a fine of $20 M.

- **GDPR Fines (2020-2022):** British Airways and Marriott hotels were charged with **General Data Protection Regulation (GDPR)**-related fines to the tune of hundreds of millions of dollars corresponding to the breach of **personally identifiable information (PII)** and sensitive data belonging to their customers. In 2021-2022, **Commission nationale de l'informatique et de libertés (CNIL)**, the French Data Protection Agency, investigated and found that Facebook and Google violated GDPR by making cookies far easier to accept than to reject for users. They fined Facebook $150 M and Google $60 M for this.

Having seen the timeline of events that made governments and regulatory bodies take notice, in the next section, we will look at the timeline of some of the major regulatory changes and announcements concerning risk.

A timeline of the top regulatory responses to financial and digital risks for stakeholders

The regulatory posture taken by governments and regulatory bodies has had to account for multiple inputs. The event timeline in the previous section should not be seen as the only prompt for some of these regulations but rather as a diagnostic of the malaise that the regulators felt compelled to address. Here is a brief overview of regulatory steps of importance in the last two decades:

1. **The Sarbanes-Oxley Act of 2002 (SOX)**: The SOX Act is a law that was passed by the US Congress in July 2002. It aims to provide protection to a company's investors from misleading and fraudulent financial reporting by the company. This act was the US regulatory response to a series of financial scandals in the early 2000's that we highlighted in the previous section – *Enron*, *Worldcom*, et al. The act mandates strict rules for accountants, auditors, and corporate officers while also making it mandatory for companies to adhere to strong recordkeeping practices. The act covers financial parts of a company's operations, such as internal controls and audit norms, and touches upon the norms for the **information technology** (**IT**) teams to store business-related data and records. The law brought severe monetary fines and criminal prosecution in case of violations and created a strong need for companies to examine their compliance posture.

2. **ISO 27001 – 2005**: While not exactly a government-mandated regulation, ISO 27001 can be seen as an organic response from the business world to maintain a common posture as the world of technology and digitization exposed them to multiple risks in terms of cybersecurity and information protection. The standard ISO/IEC 27001 was originally published in 2005 by the **International Organization for Standardization (ISO)** and **International Electrotechnical Commission (IEC)** and has gone through revisions in 2013 and 2018. Broadly speaking, ISO 27001 provides a set of guidelines for establishing, implementing, maintaining, and continually improving an information security management system. The companies that meet the stipulations of the standard can be certified so.

3. **NIST 800 – 2013**: The **National Institute of Standards and Technology** (**NIST**) is a non-regulatory body of the US, but its standards and guidance are issued after wide consultations with industry, government, and academic organizations. While the first set of controls came about in 2005, it has published revisions in accordance with US regulations, the latest to guide federal agencies to comply with the Federal Information Security Modernization Act of 2014. It also has a specific set of controls for non-government enterprises at large.

4. **General Data Protection Regulation (GDPR) – 2016**: With the advent of the modern digital age, the internet, and internet-led businesses, the European Union drafted one of the most famous recent regulations that impacts every business on the planet that deals with a European customer or operates within European geographies. The core target of GDPR is to guide and instruct companies to maintain integrity and transparency while dealing with the personal data of their customers and employees.

5. **California Consumer Privacy Act (CCPA) – 2018**: The CCPA is an example of a grounds-up regulation. The CCPA, as an act, came into existence through a privacy group called Californians for Consumer Privacy. The California Department of Justice accepted their language and requests while drafting the act, which provides various rights to Californian residents when dealing with any business. These rights include the right to know what personal data is being collected about them, absolute control over whether or not they want to share or sell their data, and the ability to ask a company to delete their personal data.

The real-world scenarios and the emergence of regulations together drove the awareness in the largest corporations of the need to be meticulous in managing their compliance. In the next section, we will see how compliance management consequently evolved as a discipline.

The various phases of compliance management and how COVID might impact the future

Within the larger gamut of risk management, regulatory risks pertaining to the digital, IT, and employee aspects of companies came more into focus. In the following sub-sections, we shall look at the evolution of compliance management as a discipline over the last two decades.

Phase 1 – GRC in the early 2000s

The early 2000s saw the establishment and proliferation of **governance, risk management and compliance (GRC)** as a practice. Undeniably, companies had GRC in some shape long before the 2000s, but the acronym came to life widely during the noughties. For most companies, this was necessitated by the SOX Act – the tough provisions of the act came as a shock to companies all across the world, and hence, the early GRC was focused on a set of controls and internal compliance as necessitated by the SOX Act.

Phase 2 – integrating GRC with the overall enterprise risk landscape

As SOX compliance stabilized as a function, businesses started seeing GRC as a part of their overall risk management landscape. This led to a couple of mini phases where companies first tried integrating GRC within their enterprise so that various departments could work within a common software space to manage all kinds of risks. But many of the companies realized over the years that a one-size-fits-all solution does not work for all functions and departments. This led to a rationalization where a master view of GRC was still integrated with enterprise IT, but the various departments looked at their compliances separately through best-of-breed solutions. However, this also meant a lot of investment, either in terms of direct specialist headcount or third-party expert views on compliance for each aspect of their business.

Phase 3 – compliance management – an agile, modern way of managing

For a lot of companies, their past experiences with GRC and their understanding of future compliance requirements guided them to a clear visualization of what solutions they needed. They looked for an integrated fabric of compliance management that enabled both their front and back offices to seamlessly look at regulations without the need for specialized knowledge. There is a breed of integrated compliance management solutions that use agile, modern architecture to give the whole company a common view of some of their regulatory risks, their framework adherence, and integration with their existing IT solutions to help them have a near real-time visualization of compliance across the organization.

What changed with COVID?

Just before COVID, we were already at a stage where companies from small to large were grappling with the following four things:

- The sheer number of regulations that they needed to be compliant with. By a conservative count, there are a few hundred regulations that could apply to a multinational business, and keeping track of a company's posture needs special efforts or investments. To give you a view of how many regulations just a few large countries in the world have, you can look at the following table:

List of regulations to be complied with:
Australia - ASD Essential 8
E-CFR
EU GDPR
FedRAMP High Security controls
Iowa's Personal Information Security Breach Protection
ISO 27001
ISO/IEC 27001:2013
NIST 800-53
Taiwan **Personal Data Protection Act (PDPA)**
Thailand PDPA
The Processing of Personal Data Law - Cyprus
The United Kingdom's Data Retention Act
Trade Secrets Act of The Republic of China
UAE - Federal Law No 2 of 2019 On the Use of the **Information and Communication Technology (ICT)** in Health Fields
UAE – **National Electronic Security Authority (NESA)** Information Assurance Standards
UAE Federal Decree Law Regulating the Telecommunications Sector

List of regulations to be complied with:
UK Cyber Essentials
UK Data Retention Act
United States of America Privacy Act
US **Clarifying Lawful Overseas Use of Data (CLOUD)** Act
US **Family Educational Rights and Privacy Act (FERPA)**
US **Federal Information Security Modernisation Act (FISMA)**
Utah Consumer Credit Protection Act
Utah Electronic Information or Data Privacy
Vermont - Act on Data Privacy and Consumer Protection
Victorian Protective Data Security Standards V2.0 (VPDSS 2.0)
Vietnam - Consumer Rights Protection Law
Vietnam - Law of Cybersecurity
Vietnam - Law of Network Information Security
Vietnam - Law on Information Technology
Vietnam - Law on IT
Washington DC - Consumer Security Breach Notification Standard
Wisconsin Security Breach Notification

Table 12.1 - Security regulations worldwide

With COVID and the push for accelerated digitization, you can expect more countries, sectors, and regulatory bodies to push for further regulations to protect stakeholder interests. The list of applicable regulations for any business is only going to increase from this point on.

- The consistent revision of the existing regulations by the regulatory bodies. After a lot of effort and investment, just when companies thought they had established some control over their compliance posture, they realized that a substantial number of those regulations had updated some of their norms.

  Many regulators are now re-evaluating their provisions in the post-COVID world. There are a lot of aspects of various businesses that have moved to the digital world, which means that the regulatory ambit will need to expand to cover these new functions and business scenarios. The subsequent revisions in existing regulations are only going to multiply after that.

- The struggles of facilitating multiple conversations between different kinds of stakeholders. A compliance officer, an IT officer, an auditor, and a finance manager will have their own takes on various compliances, and many companies struggle to facilitate conversations to get them to a convergent space. Post-COVID this trend is going to assume quite a few new colors. Many in-person alignments and workflows will need to move online to begin with. To add to that, the revisions in business models and processes brought about by COVID will need all stakeholders to reframe their understanding of their own compliance obligations.

- The pressure on regulators from the general public is only going to intensify. We have seen multiple recent scandals, such as the Panama papers and Pandora papers, that have increased general public skepticism regarding the ethics of the uber-rich and certain big companies. There has been a lot of financial impacts that COVID has exerted on people and businesses at large – the scrutiny of companies that have avoided negative impact or even grown in these times will be stringent as well, if not necessarily due to their practices, then because of the severe pressure on regulators and governments in general.

In *Chapter 11, An Introduction to Regulatory Risks*, we looked at speed, scale, and interdependence as the three key aspects of modern businesses that regulators are trying to stay on top of. In a post-COVID scenario, many businesses have already taken giant steps to partially or fully digitize their core business models and functions. This brings them right to the center of the speed, scale, and interdependence triangle. And those businesses will need to look at their post-COVID business realities and assess how much of their core shift impacts their compliance obligations. A **fast-moving consumer goods (FMCG)** company starting a **direct-to-consumer (D2C)** e-commerce portal moves into a whole new world of compliance obligations. For example, prior to COVID they dealt with retail and distribution partners who subsequently sold to the end consumer. In D2C, they might be dealing with a whole lot of consumer data where they will need to comply with the stipulated privacy and data protection norms. This is in addition to the financial implications of a very different cash cycle and reporting norms. You can expect such tectonic shifts in many businesses post-COVID. What will be absolutely essential is for all business owners to pause and reassess their compliance posture in the post-COVID world.

Summary

In this chapter, we saw the evolution of risk and risk management as an essential part of our civilizational story. As we progressed in terms of technology, digital and IT-related risks started getting specific attention from governments and regulators. This led to an influx of regulations governing companies' use of technology in their business, which companies need to adhere to. Compliance management has emerged as a discipline focused on this need – we looked at the phases of this emergence. In a post-COVID world, compliance management is going to become even more contingent on companies in new and more challenging ways. Lots of it will depend on how companies manage the vast swathes of data that they generate in their digital landscape. In the next chapter, we will delve deeper into that topic.

13
The Role of Data and Privacy in Risk Management

You can have data without information, but you cannot have information without data.

– Daniel Keys Moran

We are living in the information age, in the middle of an explosion of information. To expand on the preceding quote by Daniel Keys Moran, all data is not information, but all information is data. This means that the modern information explosion is a subset of an even bigger data explosion in the backdrop. In this chapter, we shall try to parse what this means for an average business or technology enthusiast or a student. We shall cover the following topics:

- Understanding what the data explosion is
- Understanding the enterprise and institutional data landscape
- What the top priority for governments and regulators is
- What the top priority for a business should be

Let us start by understanding what the term *data explosion* means.

Understanding data explosion

The explosion of data is an often-repeated concept of discussion – in our collective imagination, it seems to be a very modern phenomenon, but it has ancient origins. As a species, we are inherently curious about everything around us, within us, and beyond us. This is the primary reason why, out of all human attributes, we have perhaps valorized wisdom the most. All human beings have questions; wise human beings have answers. This wisdom has always had two components – observations and interpretations. The observant person picks up on different events, conversations, and occurrences that happen around them. In modern terminology, these accumulated observations could be clubbed as *data*. Some of these observations could be banal. However, some other observations could be interpreted to uncover insights. Or they could be correlated with different observations and patterns could be found. The interpreting, analyzing, and processing leads that data to become information and a curious person onto the path of wisdom. Most of this data and information was previously stored in human minds and, at different stages of civilization, propagated either through a teacher or mentor, through stories (audio), or later through books, scrolls, and art forms.

The invention of modern computational devices, starting from mainframes to large computers to **Personal Computers** (**PCs**), took the paradigm of data creation, storage, and processing to a new height. In the late 1980s and early 1990s, computing caught worldwide attention, though, for corporations and institutions, it started decades earlier with mainframe devices. And at the core of this was the age-old human search for wisdom. There was early proof of this as well. The ability to consume, compute, and process data at scale gave companies an unforeseen advantage. It is time to tell you the surprising story of the world's second-largest (by market cap) paint company – Asian Paints.

The paints industry is an old industry based on a few essential components – pigments, binders, additives, and liquids. From a market segment perspective, there are two primary segments – industrial and decorative. The industrial use of paints is to prevent corrosion and rusting and beautify equipment, white goods, cars, and machinery. The decorative segment is consumer-driven and where protective elements are important, but the core driver is the decorative value of paints for the house. Asian Paints is one of the oldest and most successful companies based out of India but serving the world. Both in Indian and Western markets, the decorative category is driven by consumer demand and the color palette remains the primary driver of value. Asian Paints, as of 2021, was multiple times larger in revenue and market share compared to their other competitors, and that too in an industry where the inputs are standardized, and the distribution channels are also more or less uniform across the market. So, what is the reason for their dominance? Well, long story short, among other things – a supercomputer.

In 1970, Asian Paints spent $150K or thereabouts to buy a supercomputer. To put things into context, many of India's and the world's leading R&D institutions and governments also did not own a supercomputer at that time. Asian Paints used this supercomputer to gain an understanding of consumer purchase patterns, including timing, cycles, colors, can sizes, and all possible details that could help them model a future by understanding the present more granularly. They used these insights to both shift the industry from a commoditized good to a more branded and personalized product and to sharpen their supply chains by making their prediction of future sales trends more accurate. In the 50+ years since that supercomputer purchase, Asian Paints has consistently maintained itself as a data-driven company. If you are inclined to find out more about their case, here is the link to a helpful video: `https://bit.ly/3vjI3m9`.

The supercomputer, though, was an industrial device that needed a very large space and a team of specialists to run. As more and more companies realized the importance of information and computation, we also saw the industry's leading technology players work on making a computer portable and friendly enough for an average consumer. Companies such as IBM, Apple, and Microsoft worked together with chip makers such as Intel to usher in a new era of personal computing. To put things into perspective, let us look at the following chart (based on data from `https://en.wikipedia.org/wiki/Personal_computer`):

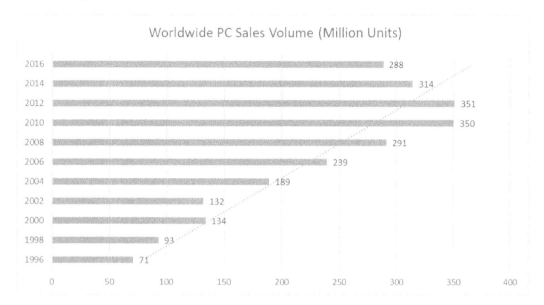

Figure 13.1 – Number of PCs sold per year

As you can see, the number of PCs sold per year grew 5x in the 16 years between 1996 and 2012. This growth will seem even starker if you go two decades back in time – the total number of PCs shipped in 1977 was 48,000. PCs made it possible to create, consume, and compute data at a large scale. If you look at the modern data explosion, a personal computation device would be the base layer of explosives. But that would still be a small fraction of the whole.

If computers caused huge growth in the creation of data, the smartphone revolution that came in the first decade of this millennium easily dwarfed it in terms of impact. The early smartphones were targeted at the enterprise market, aiming to create a category of personal digital assistants. However, they were plagued by form factors, weight, immaturity of data transfer, and wireless models. All this started changing in the early 2000s with Symbian, Windows, and I-mode platforms making the devices more trustworthy and easier to use. The seminal moment came with Steve Jobs's iconic launch of Apple's iPhone in 2007. With a capacitive touchscreen, multi-touch, easy connectivity, and design elegance, the iPhone led to a smartphone revolution. The Android operating system joined the revolution soon after and targeted the "budget-friendly" end of the market. Together, these two operating systems proliferated to billions of human beings on the planet in no time. Let us look at the numbers in the following chart (based on data from `https://www.statista.com/statistics/263437/global-smartphone-sales-to-end-users-since-2007/`):

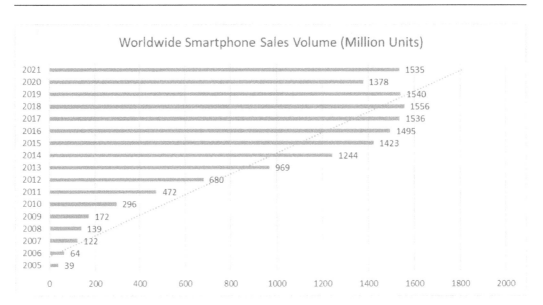

Figure 13.2 – Number of smartphones sold per year

As you can see, the number of smartphones sold per year crossed the billion mark between 2013 and 2014 and is only charting upward from there. However, these billions of smartphones, computers, and tablets wouldn't have penetrated the planet if it was not for the *internet*. The internet came to fruition with the background of the Cold War between Russia and the **United States** (**US**). When Russia launched Sputnik into space, the US Department of Defense formed the **Advanced Research Projects Agency** (**ARPA**) to boost the technological innovation drive of the US. This agency created **ARPANET**, which was supposed to help withstand a nuclear attack by helping information be routed even if the geography surrounding it was annihilated.

This created the backbone for the modern internet, but its core problem was the ways and means to facilitate computers accessing information in other computers around the world, easily and seamlessly. In 1989, Swiss physicist Tim Berners-Lee modeled the creation of a seamless network that would facilitate access to any computer from any computer. He called this system the **World Wide Web** (**WWW**) and more than three decades later, today, we can all see how this became the backbone of the modern internet. The ability to connect computers around the world and access data seamlessly came with issues. The biggest issue was that there was lots of data, and it was tough to look for any specific bit of information accurately. Yahoo first and Google later solved this by creating modern technological frameworks to navigate through the labyrinth. It is telling that both Google and Yahoo, the two most popular internet companies worldwide, started essentially as "*Search*" companies. Even when the internet was a fraction of its size today, the speed and amount of data created were the biggest pain points for users.

In the coming days and years, the internet was used to devour a large chunk of print, image, and multimedia content and revolutionize the scope of business, content, communications, and, ultimately, society and social interactions. As of today, the internet is by far the largest enabler of this data explosion. Review a few of the facts shown in the following figure to understand this better:

Did you know?

1,000

1 GB of data can ✱_
Send 350,000 emails
Stream audio for 10 hours
Upload 3,000-5,000 photos across social media

Human beings in 2020 –
Created 2.5 quintillion (10^18) data bytes daily
Did 6 billion searches daily on Google
Sent 300 billion emails daily, 2/3rd of which was spam

In 2022, the world is estimated to create 94 zettabytes of data
A zettabyte equals 1,000 exabytes
An exabyte equals 1,000 petabytes, a petabyte equals 1,000 terabytes
A zettabyte is enough storage for 30 billion 4K movies

Figure 13.3 – Some noteworthy facts about the internet

The internet has grown and become an essential part of human lives globally. In terms of digital data being created, consumed, or moved around, the implications are befuddling. The numbers associated with data on the internet have moved from mega and giga prefixes to unheard-of prefixes such as peta and zetta. The following figure tries to contextualize the exact numbers behind the *data explosion* in relatable numbers and with services that an average person uses frequently:

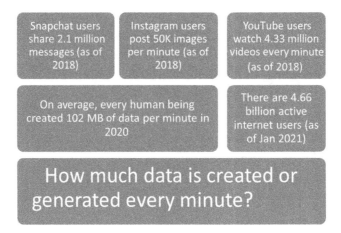

Figure 13.4 – Amount of data generated or consumed per minute

The point to note here is that some of the data points provided are slightly dated and by any reasonable guess, they would have multiplied manifold by the time you read this book. But the point is moot. An internet minute can be quantified as a unit of time in which millions of gigs of data are created and consumed. In each internet minute, an average human being is creating and consuming a whole variety of data – from entertainment to videos to e-commerce site surfing to personal messages. This data contains their most intimate secrets, humor, personal information, misinformation, disinformation, business records, transaction details, preferences, likes and dislikes, and a whole host of information that can be used or misused for eternity. In the next section, we will look at this explosion from the eyes of a company or a public institution and try to understand what it means for them.

Understanding the enterprise and institutional data landscape

In the previous section, you got a view of the end users and consumers of the world moving from the world of bulky supercomputers to PCs to smartphones powered by the internet and creating and consuming data at super speed. All the world's businesses, small and big, and institutions had to respond to this radical transformation. Some of them led this transformation by innovating their business models, some of them were born in this era and turned the dial of customer-centricity a few notches up for the whole market, and some other companies were forced to change business models and were almost annihilated before they cranked up their engines. In the middle of that, a few trends emerged across businesses and institutions. Let us look at a quick summary of these trends:

- **Get everything in real time**: When you shop on an e-commerce site, you see the real-time view of the company's inventory. You place an order; the item reduces from the company's catalog in counts, in real time. The supplier or producer of that item, internal or external, gets a nudge to produce the next batch of that item. Again, this is done in real time. The digital marketer of the company knows which products are selling more and optimizes the click journey basis in terms of both demand and availability. This is done in almost real time. The search keywords for the company's product are bid for and updated on Google, in almost real time.

- **Democratization of data**: Everyone can see the same data across the company. In the legacy world, the finance team saw and created one dataset and the marketing team created another dataset and they came to correlate once per week or even per month. With the internet, new technologies emerged that democratized data analytics. Instead of static data sources manually updated at certain frequencies, the onset of the internet also saw real-time data visualization tools coming up, making dashboards and charts available to everyone in near real time. Everyone in the company could view and bring their curiosity and thoughts to the business analytics data.

- **End of paper, endless e-records**: With the advent of the internet, paper died a slow death in the enterprise data space. From the front office employee to the C-suite, everyone created and consumed data on smartphones and PCs powered by the internet. This led to the creation of electronic records and a migration of existing data practices from paper to electronic mediums. Think of a legal case where a competitor is suing a company for patent infringement. A couple of decades ago, this legal dispute would have involved both companies scouring through their file cabinets, dusting old records, and turning up evidence for or against the claim.

- **The digital trail and the regulatory interest**: The replacement of paper records also shifted the business and regulatory need to be in control of data and the ability to scan its journey from offline to online. Governments and regulators always mandated businesses account for their data estate, especially business and customer-sensitive data. But the modernization and digitization of data made it far easier for data to be pilfered, misused, or stolen with profound implications for customers and consumers. We touched upon the details of the evolution of government and regulators' thinking in *Chapter 11, An Introduction to Regulatory Risks*, and *Chapter 12, The Evolution of Risk and Compliance Management*. Companies of the modern age are expected to have a clear digital trail of everything that is done with any specific category of data.

- **Multi-cloud, hybrid cloud, and multiple Line of Business (LOB) apps**: Cloud migration has been a focus area in multiple chapters of this book. But within that migration, what has emerged as a distinctive factor is the sheer variety of models that different companies adopted. For some companies, disaster recovery was a focus area, while for others, different public clouds perceptively delivered different qualities of service for distinct areas (such as security, speed, scalability, service trust, and others). This has led to many companies having spread their data across a variety of clouds. Add to that the fact that some highly regulated institutions and companies remain invested in captive data centers within their premises while almost all of them use multiple **Software as a Service (SaaS)** applications and on-premises applications for a variety of needs, from enterprise resource planning to customer relationship management and LOB applications. For example, many companies store data on their premises offline and create LOB applications to use that data for custom workflows inside their organization. This could be a real estate company using customer insights for the sales team or a bank using customer transaction data for internal reporting. This has led to a truly fragmented data landscape, which might become a point of concern if not handled adeptly.

In keeping with the trends mentioned previously, the modern enterprise data landscape consists primarily of the following components. Let us delve into each of them to understand what they stand for:

- **SaaS and LOB apps**: SaaS is a recent popular model that helps companies and consumers use a product without developing, hosting, or maintaining it. This product could be a chat tool or a planning tool or an intensive computing application. All you need to get started is legitimate credentials and a browser or an app that provides the end user with an interface. SaaS vendors host multiple tenants in the same cloud, and it helps them scale and maintain the products at the backend seamlessly. LOB apps can be SaaS as well as home-developed applications that a company hosts in a public cloud or on its infrastructure. The data generated by SaaS applications can be either structured or unstructured but in many cases, these applications generate unstructured data such as invoices, text documents, and custom-format reports.

- **Databases**: In the enterprise world, databases are the oldest and perhaps most transformed part of the data repository. They hold something called *structured data*, which is different from a lot of other content in different file formats such as Word docs, PowerPoint, or PDFs, which is *unstructured*. Databases emerged together with the first phase of computers, to store and retrieve information in well-defined schemas (or *structures*). As the complexity of the digital world increased, databases evolved from flat files to hierarchical databases to the modern world of relational and object-oriented databases. From the initial objective of storing and retrieving information, databases have now evolved as key business drivers by facilitating a humongous scale of storage, a superfast speed of retrieval, and adding a layer of analytics and intelligence to drive business.

- **Emails and chat**: We could have clubbed emails/chat together with SaaS applications, but the sheer ubiquity and the size of information stored on modern emails/chats have warranted this to be a separate bullet. Emails have been at the core of information creation and exchange across companies and institutions. Post-COVID, we have seen chat services such as Teams, Slack, and others replace a lot of offline and in-person channels. As an example, a company director might send a business proposal to their CEO over a chat application and get approval for it in the reply. The data that's exchanged is crucial for years to come and it lies in a chat server in an unstructured format. In a few years, a regulator might ask the company for the same details for an audit, and retrieving that from the chat application would be tough if the company has not been proactive and ensured mechanisms for that. And the regulatory need is not the only lens. A lot of institutional knowledge lies spread over chat and email. A modern company must be proactive to harness that.

- **File and network servers**: A file server (for ease of understanding, we will ignore the nuances for now) is constituted by file shares. A file share can be broken into component folders. Each folder can contain many files of various formats. With cloud migration, many companies have migrated their file servers to a SaaS service or public cloud storage. But many companies remain that store or maintain their past stores of data in file and network shares. These tools came before the advent of the cloud and were used to facilitate collaboration and role-based access to files among multiple employees. They have also been found to be potentially risky from a data security perspective and depending on the design or network bandwidth, they can be slow to access.

- **Endpoints**: A lot of data is created, exchanged, and stored on the computers and smartphones of employees. Endpoints are the important and dominant interfaces between an employee and an organization's resources. But a lot of unstructured data lies hidden from the management view on many endpoints. The end of life of a device, employee movement, and device changes obscure the data lying on these devices even further.

- **Infrastructure and storage**: Apart from standalone data storage devices or servers, many companies store their data on the public cloud too. Each of these systems can help or prevent efficiency. In most cases, standalone data storage devices run the risk of creating siloes and duplicate information systems.

A holistic look at the data landscape of a modern enterprise will tell you that it is highly fragmented and an amalgamation of legacy and modern systems. It is important to note that the journey of data can span multiple systems. For example, an invoice can be generated in a SaaS finance service, sent over email to a vendor, and a copy of that can be stored in an endpoint or a storage device. A lot of data that's generated has time-limited or no future value while other data has long-term business value. For an organization to be in control of its data landscape and drive efficiency and innovation, it needs to be very intentional about tracking, identifying, retaining, and deleting data.

What is the top priority for governments and regulators?

Governments and regulators around the world are at different stages of evolution in their journey to understand and secure their citizens. But the underlying impetus remains the same overall. We will delve into specific priorities as spelled out by the most evolved governments and regulators. But to begin with, let us try and understand how regulators and governments look at the world and what they try to achieve when they pass a regulation. The primary objective of regulations is to safeguard the interests of its citizens, its businesses, and its marketplace. These can be geography or industry-specific. The aims of regulations are both macroeconomic and microeconomic. They aim to keep their geographical economy in prime health, and they also aim to enable businesses to compete globally and flexibly. They aim to create trust for the products and services created in their industry and geography and ensure consistent economic outputs. But there is a flip side to it. Too much regulation can impede the growth of business and can create excessive restrictions that stifle innovation. All regulators are generally conscious of this and try to balance out the equation in a manner that a friendly environment is provided to businesses while still maintaining the safeguards.

In the same spirit, regulators and governments around the world have seen this data explosion and tried to ensure that this data can be effectively utilized for the betterment of the socio-economic environment while avoiding harm. In general, regulators around the world look at data from three lenses:

- **Data security**: There is a lot of concern around data theft and unauthorized exfiltration of critical data. Governments and citizens are key stakeholders when it comes to a wide variety of sensitive data. In the wrong hands, they can lead to business loss leading to economic instability, job loss leading to social costs, or even harm to national or industrial interests. Hence, regulators have mandated businesses to have a data security posture and conduct audits to ensure compliance.

- **Data privacy**: With the proliferation of the internet and internet-based businesses, as well as economic and social transactions, there has been a huge emphasis by governments around the world on the *right to privacy*. Governments want to ensure that their citizens are in absolute control of their private data and can access or delete all their private data from any company's data stores. Just to understand how widespread the recognition of this right is, let us look at this **United Nations Conference on Trade and Development (UNCTAD)** graphic:

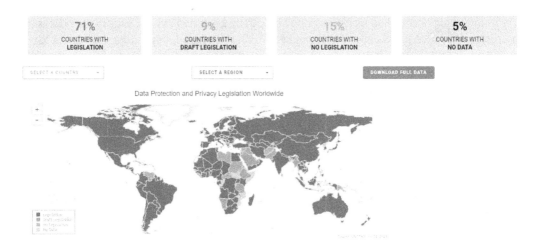

Figure 13.5 – Privacy-related legislations across the world

Let us also look at the European Union, which was one of the leaders in regulating privacy. Quoting from their official website, this passage delves into their thought regarding privacy and the benefits they intend to drive through their regulation:

"What are the benefits of EU-GDPR (European union general data protection regulation) for citizens?"

The reform provides tools for gaining control of personal data, the protection of which is a fundamental right in the European Union. The data protection reform will strengthen citizens' rights and build trust.

9 out of 10 Europeans have expressed concern about mobile apps collecting their data without their consent, and 7 out of 10 worry about the potential use that companies may make of the information disclosed. The new rules address these concerns:

- *A "right to be forgotten": When an individual no longer wants her/his data to be processed, and if there are no legitimate grounds for retaining it, the data will be deleted. This is about protecting the privacy of individuals, not about erasing past events or restricting the freedom of the press.*

- *Easier access to your data: Individuals will have more information on how their data is processed and this information should be available clearly and understandably. A right to data portability will make it easier for individuals to transmit personal data between service providers.*

- *The right to know when your data has been hacked: Companies and organizations must notify the national supervisory authority of data breaches that put individuals at risk and communicate to the data subject all high-risk breaches as soon as possible so that users can take appropriate measures.*

- *Data protection by design and by default: "Data protection by design" and "Data protection by default" are now essential elements in EU data protection rules. Data protection safeguards will be built into products and services from the earliest stage of development, and privacy-friendly default settings will be the norm – for example, on social networks or mobile apps.*

- **Data retention**: Many industries and regulations expect companies to retain specific kinds of data for a specific amount of time and then delete or dispose of it. Seen from the regulators' perspective, company-consumer transaction data (including communications around it) should remain available for scrutiny in case of a future dispute. Similarly, many public sector entities create or store data of public interest and governments warrant them to preserve it for long durations and then move it to central archives. Many other scenarios mandate companies retain data and keep visibility of everything that happens around the data (access, editing, movement, and so on). There is an additional business imperative too: a large chunk of data created or captured by a company does not have any business value. It lies unattended, without any use in the company's stores, creating storage costs and impacting efficiency in data operations. Regulators want companies to be intentional about both retaining business-critical data and deleting personal or non-useful data at stipulated timelines.

Regulatory logic emerges from a variety of concerns and goals. In this section, we summarized those in three buckets. But that does not mean that this is an exhaustive list. The preceding summary, however, captures a decent majority of those concerns and goals. Now, let us look at how companies should prioritize their asks from their data estate.

What should be the top priority for businesses?

In the world of data glut, with buzzwords such as artificial intelligence, machine learning, and data science, the priority of businesses has emerged as tiered yet simple. Different companies with different amounts of data and dependent data flows might want to approach their data estate differently. There are several leading technology-based companies where processing all their existing data and innovating on top is the order of the day. There also are quite a few companies whose touchpoints are predominantly physical, and the data footprint is not enormous. In this section, we will look at the average use cases and prioritize them for most businesses around the world. With that in mind, here are a few things that apply to a broad spectrum of companies:

- **Awareness of data repositories and impetus on data quality**: A company has several tools and repositories for data creation and storage. At the very minimum, a business should have visibility of all these locations. Moving beyond, it is important to map the data flow between multiple systems and keep mechanisms to ensure that the data does not lose fidelity or quality at any point. End-to-end data life cycle management is a goal that every business should have.

- **Intentional retention and deletion**: While highly regulated companies in industries such as banking and financial services are likely to be maintaining a retention schedule, it is equally important for all companies to be very proactive and intentional in retaining or deleting data. There is a sense of anxiety involved in most companies when it comes to deleting data. Almost everybody fears deleting something that can be of consequence later. But information governance and data life cycle management is an evolved field and companies need to refer to models such as the IGRM framework (`https://bit.ly/3kscVKX`) and create cross-functional teams to start the conversation and align on goals and methods.

- **Drive operational efficiency**: Data remains at the core of a company's operations. Companies need to look at data flows from the aspect of efficiency. For example, **Internet of Things** (**IoT**) technology brings a lot of operational data from hardware and equipment into the manufacturing sector. This data can be directly used to increase asset lifespan and employee productivity. Also, data in itself is a cost driver in terms of storage or compute costs. Having a conscious data rationalization strategy can increase efficacy for organizations.

- **Managing data risks**: We expanded on data risks in *Chapter 3*, *Visible and Invisible Risks*, *Chapter 6*, *The Human Risk at the Workplace*, *Chapter 7*, *Modern Collaboration and Risk Amplification*, *Chapter 8*, *Insider Risk and Impact*, and *Chapter 12*, *The Evolution of Risk and Compliance Management*. It should be a top priority for businesses to be proactive in mitigating data risks.

- **Meet regulatory requirements**: As explained earlier in this chapter, regulators are coming up with ever-evolving guidance and regulations related to data. By one estimate, 220+ updates are coming per day from thousands of regulations around the world and a large enterprise needs to be up to date with all these requirements. Investment in compliance and regulatory management is the order of the day, either through in-house resources, tools, or third-party help.

While this list gives a beginner's view of priorities, companies in highly regulated verticals such as banking or healthcare will need to do much more than just this. However, we can conclude that, for all of them, the points mentioned here will still be applicable.

Summary

In this chapter, we looked at the emergent data revolution and checked various historical perspectives. We looked at personal computing, smartphones, and the internet as the three top drivers of this revolution. From a business and institutional perspective, these three drivers created unique opportunities and challenges. We saw how Asian Paints took the lead in analyzing and understanding data to drive business growth 50+ years ago. However, emergent patterns caused concerns to various governments and regulators. Due to this, we provided a summary of their viewpoint on data explosion. We finished this chapter by understanding the basics that a business needs to keep in mind to survive and thrive in the throes of this data explosion.

We will now change gears from data and privacy risks and try to put on our predictive lenses to look at some trends for the future in the next section of this book. We will cover the trends of remote working in the next chapter.

Part 3: The Future

This part focuses on the future of work and technology. It covers remote work and the virtual workforce and explores the concepts of automation and virtual humans. Additionally, this part delves into the impact of AI on future events, particularly with regard to lockdowns. The aim of this part of the book is to provide insight into how technology will shape the way we work and interact with one another in the future.

This part of the book contains the following chapters:

14
Remote Work and the Virtual Workforce

In this chapter, I will discuss my perspective on remote working. While the COVID-19 pandemic has temporarily altered the way we work, I believe that it is **artificial intelligence (AI)** that is fundamentally changing the nature of work, including where and how we do it. The concept of work is being redefined by AI, and it will be interesting to see if it ultimately matters where we are physically located. Additionally, the pace at which this new type of work is being adopted is remarkable. We will also see the emergence of new categories of workers, which I refer to as *work beings*. Overall, I am more concerned about the impact this shift will have on human social connections, emotional and mental well-being, and the role of creativity in work.

In this chapter, we will ask the following questions:

- Will remote working be a permanent change?
- Do I have to work alone when working remotely?

Will remote working be a permanent change?

It's important to understand how, when, and where work has changed. We transformed our kitchen tables into workstations, and kids' video classes dominate our homes. The line between work and personal life has blurred. Will we go to the office like in the good old days or is working remotely forever?

Some companies have said they will stay working remotely permanently, while others are looking to call back employees to work in the office. Most companies are exploring a hybrid approach, but what is the best option for a business and its employees?

The way we think about work is evolving and leading to a revolution in the workplace. At my company, with 200,000+ employees globally, our CEO Satya Nadella's vision for the future of work aligns with the idea of a hybrid approach.

Using this approach, it can be said that there are a few common questions about remote working that almost every business leader needs to ask, such as the following:

- What kind of safety measures should the business adopt to ensure employee safety and work continuity?

- Will employee productivity increase, decrease, or stay the same?

- It's good to save on office infrastructure costs, but is this sustainable?

- What happens to my organization's culture?

- What are the risks to my organization in this model?

In my opinion, a lot is changing in the work sphere than what meets the eye.

Scope of our work

Work itself is changing at a fast rate, and the revolution impacting that is AI. Once we have COVID-19 in our rearview mirror, we will realize that the bigger revolution was how AI came and changed our work culture. While the trend of remote working and where we work from is capturing everyone's attention, what many of us miss is that the core nature of work itself is evolving way faster than the places we choose to work from. The evolution in the work process is making where we work less important. We can also refer to this as what we work on.

I remember when I used to write software programs – my manager always asked me to review the code of my peers. It was difficult reading and correcting someone else's code instead of rewriting it all myself. I always wanted to sit down with the person first to understand the logical construct before I started reviewing the code. Today, we have AI self-learning algorithms, such as the one developed and researched by Stanford University, which reviews code and fixes issues within it. This changes the working method, with the review no longer being performed by a human (`https://ai.stanford.edu/blog/DrRepair/`):

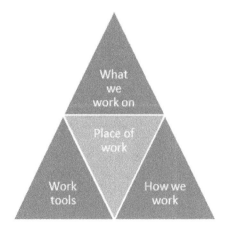

Figure 14.1 – The work context in the digital world

Let's take another example, where documents such as a scope of work or legal contracts needed to be reviewed. In the old days, professionals and subject matter experts used post-it notes or thick sketch pens (as shown in *Figure 14.2*) to highlight the areas to be reviewed or changed. Then, the era of software such as Microsoft Word Online and Google Docs introduced co-authoring, which enabled multiple people to view, edit, and track changes in real time. It also had some useful features, such as being able to leave comments for review and tag people for their attention. Now, review work has changed even further, as we can use AI to review much better than even the smartest human.

The progression of technology has led us from sticky notes to word processing and now to AI-generated content checkers, as shown in the following figure:

Figure 14.2 – The evolution from physical tools to digital tools

In a case from Lawgeex, an AI model reviewed five legal documents as well as some of the best lawyers and outperformed every law team, both in speed and quality. Specifically, the AI model had an accuracy rating of 94%, while the lawyers achieved an average of 85%.

This is a good segue into the virtual workforce part of this chapter's title. Each era of technological advancement has introduced automation in the work arena, challenging established human methods, and, many times, replacing them completely.

What intrigues me more is that the decade-old practice of having a pen for marking paper documents evolved into the digital practice of reviewing documents with comments, tags, and co-authoring in just 3-5 years. It's surprising how quickly users adapt to AI-based document checks, from legal checks to authoring checks, once they become aware of this option. The nature of work has shifted, with many tasks that were previously completed by humans now being reviewed for completeness using AI technology.

Work tools

In addition to work itself, work tools have changed. The modern tools that operate inside our phones and laptops are bringing new ways of working and making our places of work irrelevant. For people whose careers extend beyond 20-30 years, the days of desktop computers and phones with coiled cords must still be fresh in their memory. I still remember the big Nortel Meridien phone we used to have with a speaker for conference calls. I used to struggle to remember the procedure to bring multiple people into a conference call. Today, we have our work calendars full of virtual video calls, and we don't even worry about adding multiple people to video calls; it happens with a click. Our working tools have transformed and continue to get transformed further. Today, we enter a Teams/Zoom call with our video, and soon, we will step into the Teams metaverse, where our 3D avatar will mimic our behavior, giving the illusion of our presence. While on a video call, I need to switch my camera on; in the metaverse, that requirement will be optional. So, I could be in my kitchen making my favorite dish, while the metaverse shows me in a board room talking in a business suit. If I can present from my kitchen, why can't I work remotely or from a different location using these new tools?

Advancements in technology have taken us from traditional telephones, to automation in the manufacturing process and now to augmented reality, as shown in the following figure:

Figure 14.3 – The evolution of industrial tools

Let's now look into how the way we work will be impacted by different tools based on the trends so far.

How we work

It's easy to visualize remote working for jobs that can be done using home computers. We can call such workers knowledge or information workers. Is remote work even applicable if the job is to paint a car or work on the assembly line of a car manufacturing company? The nature of the job would require you to go to the car manufacturing plant or the factory for work.

Yes, it's difficult to visualize how you could paint a car from a remote location. However, what if the car painting process also gets digitized with the introduction of robotics? In short, what if you could operate a paint machine remotely, one that has robotic arms that can paint, instead of you manually doing the task standing next to the car? Then, you could be miles away and paint the car. Yes, someone needs to be in the factory to fill the robotic arm with paint, switch on the robot, and initiate a few prerequisites for automation to work, but you could still do your job remotely.

Today, robots can build cars automatically. Automation technology has advanced to the point where robots can handle various tasks in the car manufacturing process, such as welding, painting, and assembly as shown in the following figure. Human workers are still involved in overseeing the production process and ensuring the quality of the final product. What we need to assess is whether humans can work from home instead of being physically in the factory (you can find the original figure at `https://roboticsandautomationnews.com/2020/02/24/special-report-robotics-and-automation-in-automotive-manufacturing-and-supply-chains/30374/`):

Figure 14.4 – Scope of work and tools

My point is simple – our work, the tools we work with, and how we work are changing faster than we can imagine. This digitization tsunami, with all the benefits of AI, automation, RPA, and machine learning, is bringing a transition that is here to stay. From a work perspective, this change is making it more feasible to be delivered remotely.

I feel that in a few years, some jobs shall permanently transition to remote working. Take, for example, the outsourcing call center industry, specifically call centers. Large employee hubs in metropolitan areas have mostly, by design, changed to a remote caller-based model. The pandemic has taught us that just like a caller that can call from anywhere in the world to a call center, the receiver can also be located anywhere to answer the call. Avatar technology is an emerging field – it's an advanced form of digital representation of a human being that can be used in a variety of contexts, such as virtual reality, remote working, and customer service. As technology continues to evolve, avatars could become what I call work beings in certain industries. For example, in customer service, work beings could be used to interact with customers in a more personal and efficient way. In remote working, work beings could be used to represent employees in virtual meetings and other work-related activities. One question that I find most companies and managers need to answer is what systems they will put in place for remote workers to use a human face-to-screen avatar. Some trends began to emerge at the end of 2021. The 2022 Microsoft New Future of Work study found that "*employers are still embracing hybrid less than employees want: As of March 2022, US employees still want to work from home more than employers are planning to allow (around 0.5 days/week, depending on the type of worker) (Barreroet al. 2022). Globally, we see the same dynamic: In AIPAC, 40% of employers are planning flexible work in 2022, but 60% of employees want it.*" For more details, you can download the report from `bit.ly/3SkVVVm`.

Having looked at the past and possible future of remote work trends, it is now time for us to look at the psychological and physical aspects of remote working for an employee.

Do I have to work alone when working remotely?

One frequent occurrence for me is when I am at work in my office, working on an important task, and suddenly, I get a new chat alert. I quickly click on the chat software to see a ping from my colleague, asking whether I am going to join a virtual overseas call that evening. I type a response, and the conversation takes a couple of minutes to wrap up. In the meantime, the important task I was working on has receded from my focus. I am sure this will resonate with some of you. According to the University of California, Irvine, their employees get interrupted once every 11 minutes on average. Each interruption results in roughly 25 minutes of lost productivity, as employees take some time to return to tasks with the same level of intensity that they were working at before they heard that instant messenger alert. Sometimes, if it's not even an alert; it's the human urge to quickly open a browser and check social media news, or browse some other topic of interest. I define the work that we do today as a collection of micro-tasks that we accomplish during a certain period of the day. It is an emerging area of concern for organizations that need dedicated deep work from their employees – specifically, how the reorientation of an individual's approach to micro-tasking shapes their ability to concentrate on one task for sustained a period.

Whether you work remotely or from the office, micro-tasking has become ubiquitous. We can link it to the emergence of the internet and its myriad of content, coupled with consumer tools and software targeted at individuals and their life trends. Social media, the shopping list on our favorite e-commerce sites, news, and interests – these are just a few of the things that many people have in the back of their minds, whether they are in the office or working remotely. I am sure that with the prevalence of remote work, the affinity for micro-tasking is set to increase.

It's unclear yet whether this is bad for individual or institutional productivity and outcomes. I subscribe to the idea that multitasking can be beneficial when tackling multiple tasks. Working on several smaller tasks simultaneously allows my mind to take a break, change perspective, and ultimately complete the overall job more efficiently throughout the day. I don't feel the need to rush and finish one task before moving on to the next. If I have planned to accomplish five tasks in a day, it does not necessarily matter in what order I complete them.

The future workforce would be quite different. In addition to human beings, the future workforce will have new *work beings*. New work beings? Yes, there will be bots, robotic process automation or automation workflows, avatars, and metaverse entities that will be empowered to interact with you. If we think about a fully automated car factory, a car would be manufactured in the assembly line with almost no human intervention. Let's say the car will be assembled in the assembly line that goes by the name Jackie. Let's say it will be painted by an entity named Pacman in a workflow without any human intervention. Both Jackie and Pacman are AI-powered workflow engines that are overseen by just one human operator. Both Jackie and Pacman have replaced a few hundred workers who were working on the shop floor a few years back. Now, we only have three to four operators and a bunch of programmers who keep updating the program that dictates what needs to be done by Jackie and Pacman.

So, you are not working alone. The operator may be sitting miles away from the actual factory. Jackie and Pacman are AI-based work beings that ensure the cars keep getting manufactured. When they need some help, which might be regarding issues such as the paint tank running low on paint, the temperature of the plant not being ambient for manufacturing, or any other unexpected challenge that has either input-cost implications or has the potential to pause or slow down the manufacturing assembly line, both Jackie and Pacman can automatically ping the operator.

The human operator must understand how best to resolve the open challenge and get both Jackie and Pacman to start working on the assembly line to their fullest capacity.

You may be asking, "So, you mean to say that a non-human work being will ping me?" What I mean is that you will get a ping from AI software. The ping will tell you that there is a problem and you need to resolve it so that manufacturing can continue. This is already a common procedure for most of us. For example, you may have Uber or an equivalent app installed on your phone. You will get a notification from an AI algorithm owned by Uber when your cab has arrived, and then you step out of your home, office, or cafe toward the cab. We already receive pings and notifications from many apps and services telling us to act in the physical world. This will become common in most organizations as AI algorithms, automation, and **robotic process automation** (**RPA**) become more ingrained in business processes.

Working remotely is going to get more interesting. Today, you receive instant messaging in a very 2D, non-graphic format. Eventually, we will wave goodbye to dull 2D video conferencing and instant pings. A 2D video call is a traditional video call that is typically done using a webcam and a flatscreen. The image is two-dimensional and can appear lifelike. A 3D video call or holographic call, on the other hand, uses technology to create a more immersive and realistic experience by adding a third dimension to the image. This can make the call feel more like an in-person conversation and can improve engagement. Holographic technology is coming, along with 3D images, holographic avatars, **virtual reality** (**VR**), and **augmented reality** (**AR**):

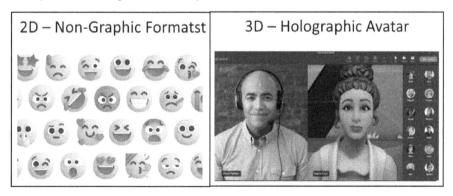

Figure 14.5 – The effect of visual elements on engagement

The preceding figures can be found at www.windowscentral.com/microsoft-clarifies-where-its-new-2d-and-3d-fluent-emojis-will-appear-windows-11 and www.microsoft.com/en-us/mesh.

Working with Microsoft's HoloLens 2 headset and applications such as Spatial has allowed remote workers to use avatars in virtual meetings, interact with 3D projections of projects with fellow team members in real time, and drag and drop information into a collective holographic space. I am personally betting on VR and AR playing a primary role in our remote workplace life in the coming years. To illustrate this with an example, I recommend that you visit https://spatial.io/. According to Spatial's website, "*Spatial is dedicated to helping creators and brands build their own spaces in the metaverse to share culture. We empower our users to leverage their beautiful spaces to share eye-popping content, build a tight-knit community, and drive meaningful sales of their creative works and products. We also empower our users to create beautiful and functional 3D spaces that they can mint as NFTs and sell/rent to others looking to host mind-blowing experiences.*"

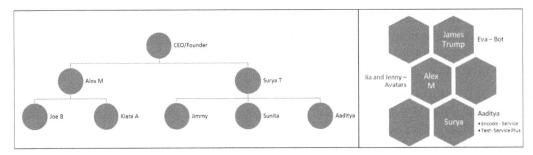

Figure 14.6 – An organization structure with humans and digital humans

In all organizations, **human-to-human** (**H2H**) interaction has been standard for decades. We have also seen numerous examples where humans operate machines and machines assist humans in accomplishing work. Some of the most common examples are buses, trains, excavation machines, microwaves, and dishwashers.

We are now in an era where machines operate and assist humans. Google Maps is a simple example of a machine existing for human beings who can then find the right path. Chatbots also fall into this category. We are also going to see a new category altogether, which is machines assisting other machines. Imagine your personal chatbot calling your doctor as a result of your health vitals being reported by your Apple Watch.

In this scenario, one machine, which is the Apple Watch, speaks to another machine, a personal chatbot with AI. The personal chatbot realizes, based on the vitals given by the Apple Watch, that it needs to call the doctor. You can easily imagine what would happen next. It is possible that in the future, an Apple Watch or other similar device could be programmed to detect abnormal vital signs and contact a doctor or medical professional automatically; the idea here is to make you aware of new ways of interacting between humans and machines, and a mesh between these entities. In 2018, *Harvard Business Review* published an article about how collaboration between humans and AI can strengthen each of them, and we are already some way down that road.

You can find multiple articles sharing the future of devices at www.hbr.org. As mentioned previously, you are never really alone in the workforce. I think it might be fun to have an AI waking you up with a nice "good morning" greeting every time you enter a virtual metaverse call, don't you think?

Summary

A lot of what we've covered in this chapter doesn't look like science fiction anymore. Technology may create headaches and disrupt productivity. In my opinion, the expansive domain of VR coupled with hybrid teams, a remote workforce, and sentient work beings will reduce cost, boost productivity, bring innovation, and give humans more choice and flexibility.

I also believe this heightened digitization will come with its own challenges. These challenges include digital cybersecurity risks, as well as the sustainability of operations, both ecologically and in terms of how individual humans are capable of adapting and adjusting. The challenge will be to figure out what tasks will be left for humans to do if work beings take over everything. I don't think it's realistic to imagine that everything will be built and serviced by work beings. We will still need human operators, human developers, human architects, human testers, and many other roles where there is a need for humans to intervene, own, and operate an entire system. I hope that this chapter piqued your interest in automation and AI, which we will delve into deeper in the next chapter.

15
Automation and Virtual Humans

In the previous chapters, the concepts of *human beings* and *work beings* were introduced. The term *human beings* refers to individuals who perform work tasks in a traditional sense, whereas *work beings* refers to new categories of workers that may be emerging as a result of technological advancements such as **artificial intelligence** (**AI**) and automation. These *work beings* could include things such as AI-powered robots or avatars that are capable of performing tasks that were previously done by humans. As technology continues to evolve, we will likely see more and more "work beings" in the workforce, and this could even change the degree of automation that can be used to complete work. When you're working remotely, real human beings also appear as an image on your computer screen. You can relate to them because, in your mind, you believe there is a person behind that image that you see on your computer screen. The screen presents a real-time video of the person sitting behind the camera on another computer. This person could be sitting in the room next to you or could be thousands of miles away from you. Until fairly recently, it was not possible to imagine that you could have a real-time conversation with other people and look at their video stream in real time, hear their voice in real time, and also be seen and heard by them with your video in real time.

Have a professional-looking office setup as your background. Today, most of us enter into a Teams meeting or a Zoom call with the background of our choice. Some people prefer a beach background, some people prefer an abstract background, and some prefer natural home or office images as their background. Most of my calls are official and I do see lots of office image backgrounds when people across the globe join these video calls. Sometimes, it's difficult for me to tell whether they are using an artificial background or sitting in the environment visible on my screen. As more realistic and dynamic backgrounds that include motion in them appear, soon, it may be impossible to differentiate between whether somebody is using an office image background or is sitting in a real office. Moving one notch above backgrounds, today, I can do a video call using my avatar. My avatar is also my metaverse image. As the world moves from actual videos with a virtual background to digital twins, it will become difficult to assess whether the person you are talking to on your computer screen is real or virtual. In this chapter, we will further explore how automation and AI will create different varieties of work beings that may not be real humans and will work alongside real humans as companions, admins, or sometimes even managers leveraging automation.

In this chapter, we will cover the following topics:

- Automation in this digital age
- The maturity of chatbots
- Digital humans
- Digital humanoids

Automation in this digital age

Automation simply means that a machine will automatically do the job for you. Today, a classic example of automation is the washing machine or the dishwasher in your house that automatically does the job of washing your clothes or cleaning your utensils. As input, you either need to provide dirty clothes or utensils that you need the machine to clean and it cleans them for you. Talking in terms of business, automation comes in various forms and capacities. Automation in business is also popularly referred to as **robotic process automation (RPA)**.

RPA helps businesses automate tasks that are repetitive and can be performed easily by machines. For example, you can automate responses to customer calls, and you can automate the manufacturing of phones, but you cannot automate which colors should be painted on cars based on the booking that has been fed into the system.

What automation does is reduce human intervention in processes. This intervention is reduced by a predetermined decision that is programmed into the machine. Current machine learning technologies promise to take this one step further eventually by also involving machines that will be able to predetermine decisions supposed to be taken by humans or other machines. The autonomy we give to our machine and the level of automation will determine the impact it has on business processes.

Now, it is easy to visualize how manufacturing products under the control of computers are an example of industrial automation using computers and AI. When cars are manufactured in assembly lines with machines and robots, they also fall into this category. When you extend this automation to the work that humans do, such as taking orders, answering queries on the phone (in the call center industry), or skimming through thousands of resumes that have come for an open position, and then these kinds of repetitive work that used to be done by skilled humans are also done by skilled machines, we can say that a new dawn of automation has arrived. So, am I saying that the work that used to be done by skilled humans would now be done by a skilled machine?

New-age companies are those that use cloud computing and AI as their first choices. These new-age companies are on a mission to increase productivity using computing and are coming up with products and services to deliver automation. Some of the leaders in this segment include Microsoft with Power Automate, UI Path, and Automation Anywhere, just to name a few. We are familiar with replying to emails. The following screenshot (the original image can be found at `https://powerautomate.microsoft.com/en-us/`) gives a glimpse into AI-based automation software that can even reply to emails on your behalf based on your habits and on approved rules about what to reply:

Figure 15.1 – Modern email automation software

If you look at the preceding screenshot of Microsoft Power Automate, I was intrigued by this graphical interface and its ease of use, with which I can set up automation and workflows. What also excited me was that in this Power Automate framework, there were thousands of pre-built templates in the library that I could use – for example, a template that could be used to automate email replies from support email IDs provided by most companies to their customers. Email addresses can automated or play a role in automation in a variety of ways. One of the most common uses is for sending automated emails. For example, an e-commerce website can use automation to send automated emails to customers with their purchase confirmations, tracking information, and other relevant information. Another use of email addresses in automation is for sending automated reminders; for example, for an appointment, a task, or a due date.

Additionally, email addresses can be used as a way to trigger an automated process or workflow. For example, an email address can be set up to automatically forward emails to different individuals or teams based on the content of the email. Or an email address can be set up to automatically trigger a script or program to perform a specific task, such as updating a database or generating a report.

Inbuilt templates in Power Automate help you create automated workflows so that humans don't have to respond to the queries that correspond to support email IDs. What you can do with any automation framework is truly limited only by human imagination. I think we're not far from software itself being able to read, listen to, or observe business transactions and respond to requests from customers across email, telephone, and social media, and driving the use of automation for increased efficiency and productivity.

The maturity of chatbots

I believe that in the upcoming years, regardless of whether we work remotely or in an office, we will have more work beings with us. When we look back in a few years, we will realize that chatbots were among the first work beings in the workplace. Today, most of us have interacted with chatbots. Today, I communicate more using messages than my voice, and more via the written word than my voice in general. I send WhatsApp messages to convey my point rather than calling people and talking to them. It's not a coincidence that chatbots are becoming more and more popular because they offer similar freedom and flexibility to messaging.

Chatbots will mature over time. A good chatbot will provide short and high-value interactions with customers through tasks that are automated using automated workflows. An amateur chatbot will be available 24/7, offering customers the ability to interact on demand anywhere, on any platform or device. Amateur chatbots will also gravitate toward personalization. Chatbots will start conversations with context, and they'll talk to you in a natural way as humans do. A mature chatbot ecosystem will also promote a company's products and services across email, telephone, and social networks on which a company's prospective customers are already engaged. As chatbots become smarter, they will improve the customer experience and reduce the need for human interaction.

Not only do chatbots help customers but they will also bring in value for employees. Chatbots will increase productivity by letting them focus on more critical work. They will also allow your employees to interact with customers on any platform on any device. These intelligent bots will become virtual secretaries so that employees can engage with chatbots as if they were talking or chatting with humans for simple activities such as checking their job goals, company benefits, or their amount of remaining paid leave. This will increase employee satisfaction and reduce fatigue when dealing with high-volume repetitive tasks with fellow humans.

I've categorized chatbots based on their maturity into three levels. Level 1 chatbots mostly answer basic questions by going through a document or top FAQs. These chatbots can easily answer questions such as where your office is and about your company, products, product features, prices, and so on. I would also add capabilities such as meeting management, calendaring for appointments, or helping people to park their cars in available parking slots under level 1:

Maturity Level – 1	Maturity Level – 2	Maturity Level – 3
<ul><li>Answer basic FAQs</li><li>Understand language contexts</li><li>Speech/voice interactions</li><li>Compliance with regulatory requirements</li></ul>	<ul><li>Ability to use cognitive features, such as the ability to read images and locations using images from a camera or location feed from a phone</li><li>Understand emotions and then generate/change/switch contexts</li><li>Speaker recognition for different automation flows</li><li>Translator services across languages</li><li>LOB/API integration</li></ul>	<ul><li>Integrated with custom devices with persistent memory of its last conversation</li><li>Use of machine learning to self-adapt and respond</li><li>Custom speech, vision, and response services</li><li>Avatar body and shape to make it a human-like experience</li></ul>

Table 15.1 – Maturity level for bot development

Maturity level 2 chatbots, on the other hand, in addition to level 1 capabilities, can go into line-of-business applications and generate transactions. An example of a maturity level 2 chatbot would not just tell you about a product or service and its price but also provide a payment link. A maturity level 2 chatbot, after sharing a payment link, will also follow up if you have not made the payment. Once you have made the payment, the maturity level 2 chatbot will also update the product inventory to deduct the purchased item. It will also send a notification to the customer and track the shipment if configured to do so. The chatbot can also dispatch the product or service at a convenient time for you. The chatbot will also update the reporting structure. The entire business transaction is carried out by the chatbot without any human interface, support, or interaction. Let's take an example at this maturity level where you want to renew your car insurance. You interact with a chatbot that can pull all the details of your current insurance by going to the web services of your current insurance provider company. The chatbot can also scan offers from other insurance providers for your car. The chatbot then goes to your car manufacturer to see whether you have made any claims and validates the information with the car manufacturer. Using this information, the bot gives you access to proposals from various insurance companies. You select the best insurance policy from the list and the bot presents you with a payment link. You pay on the payment link and the bot shows you your new policy. The bot then also communicates with the insurance company, updating its records with the new policy or renewal of your old policy. The bot does what a call center would have done otherwise. The following diagram depicts the differentiated maturity of chatbots:

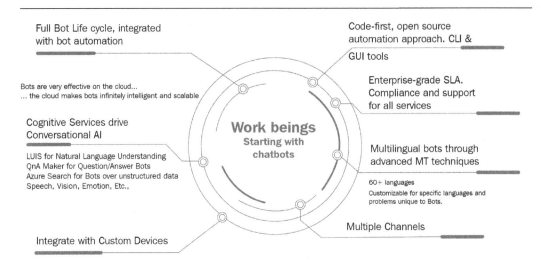

Figure 15.2 – Chatbots as new work beings

A maturity level 2 bot can facilitate a transaction and has the cognitive capabilities to scan the user's environment and move a human-like conversation forward, but a maturity level 3 bot can do way more than what humans today can do. Imagine that you can talk to a political leader who is no longer alive and understand their perspective on the world. Imagine talking to Einstein to learn about physics when we all know that Einstein died a long time ago – well, what I mean is talking to a digital avatar of an influential person who may not be alive anymore. The idea is to create a metaverse in which you're able to see a human-like entity, which I call a work being. Now, we see work beings with a form factor. This form factor can transform itself into a 3D digital avatar of a known personality that you admire or will listen to. A conversation with a bot that uses this form factor feels more connected, as though the bot was a real human being. To make things more realistic, what if this avatar also meets you every time with different clothing and hairstyles or perhaps adopts other human-like behavior, such as changing their spectacles? Also, if I add environmental factors such as weather, economy, and politics into the context that can be applied to the conversation by the avatar, am I truly making it like real life? So, how do I know whether this avatar is real or virtual? The reason I brought up this point is that today, companies such as Microsoft and Facebook are trying to create digital twins of real-world humans, who will be represented by 3D avatars. I can only imagine it will be very difficult to distinguish between them. Let's move on to the next section to understand digital humans and where they stand today in terms of innovation.

Digital humans

If we do a quick Google search and look at the 11th International Conference, DHM 2020, *global group* research on this topic, they had their 22nd international conference, named *HCII 2020*, in Copenhagen, Denmark from July 19 to 24, 2020. Reading and interpreting the summary notes will give you a glimpse into what work this group is doing in digital human modeling in health, safety, ergonomics, and risk management.

The group focuses on how machines can use cameras to supplement a human-like ability to read emotions, read beyond facial expressions, and read body language. This group of scientists is trying to look at creating an AI model and connecting gestures to an environmental context, such as in restaurants, homes, offices, or on trains, so that software can recognize and reflect emotions, mental states, and possible opinions suited to the environments they have learned about. Let me, for simplicity, call it an AI-based gesture reading model. It's about how machines can simulate opinions and then make decisions using this AI gesture reading model. It also looks at sign language and mixes it with facial gestures so that machines can read human intent and utterances. By leveraging utterances from real-life public conversations, they can feed them into machines for them to use to respond like humans. The group shares models for mapping these utterances to hardware such as Raspberry Pi. Yes, it's all happening now.

You can observe the visual appeal of these avatars by visiting sites such as `Soulmachines.com` and `Uneeq.com`. When you start using these digital humans, you'll find that they do share, talk back, and have human expressions. I think we may see the ability to use a device camera to factor our expressions into talking points in newer iterations of this technology. A further enhancement would be to factor in our surroundings so that if I am traveling in a car, sitting in a café, or by a beach, the software will take this context into account. This will make software alive with emotions to express, comments, and appreciation of what they see, as humans have. There is ongoing research in the field of **artificial general intelligence** (**AGI**) and **artificial super intelligence** (**ASI**) that aims to create machines that can think and reason like human beings, and also have emotions. Currently, work beings do not have emotions in the traditional sense. They can simulate emotions and respond appropriately based on the situation, but they do not experience emotions as human beings do. However, as technology continues to evolve, work beings may eventually have some form of emotions. The following figure depicts the avatar for Einstein made by an organization called Uneeq (`www.uneeq.com`):

Figure 15.3 – Digital human visualization

Please use the following link for more information: `https://developer.nvidia.com/blog/an-era-of-digital-humans-pushing-the-envelope-of-photorealistic-digital-character-creation/`.

Digital humanoids

I would not say that Sophia was the first digital human robot but was surely the first to get so much attention – digital human robots... quite a heavy concept to digest. A digital human robot is a machine that looks like a human being. It's a powerful computer with arms, legs, cameras, and software to do things such as walk, take a book off a shelf, and bring you coffee so that computers that look like humans can also act like humans:

Figure 15.4 – Digital humans in the physical world

Sophia was formally switched on – or born in human terminology – on February 14, 2016. It caught my attention when Saudi Arabia gave Sophia citizenship in 2017, initiating a shocked and confused reaction from most people in the world. You could say people weren't even sure how to react at all to the news. The juggernaut did not stop, and the United Nations named Sophia its first Innovation Champion for Asia and the Pacific in 2017. Sofia is the first non-human to be given such an honor, but I am sure not the last.

Elon Musk, who was the creator of the famous autonomous electric car Tesla, innovated self-driving cars a few years back. Recently, he also announced his company Tesla's entry into humanoid technology. His next innovation slotted for release in 2023 looks straight out of the movie *iRobot*. If we look at what's under the hood, we'll find a familiar computer on a chip or an FSD computer as used in electric cars, which we can see in the following figure. It comprises a battery, like a car does, an autopilot mode, as found in a Tesla, and multidimensional cameras controlled by simulation and AI algorithms. It will be interesting as this robot comes to life to see what kind of work it can do – for example, in retail or maybe even lifting kilos of weight. If you look at the following figure, the Tesla robot (the original image is from `www.tesla.com`) looks like a human and is carefully made to resemble a human in terms of height, weight, and features such as arms and legs:

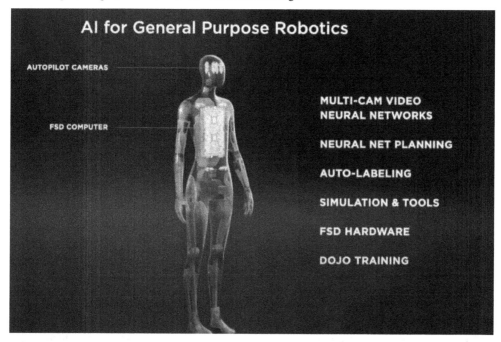

Figure 15.5 – Bots looking like humans

It's likely that once we have humanoids around us, they will take over our physical work. So, should humans even do labor-intensive physical work such as lifting weights on a shopfloor or in a factory, cutting wood, painting houses, and so on? Imagine that most of the work of blue-collar workers was done by humanoids and most of the work of white-collar workers was done by invisible AI and algorithms – what will humans do in the future? This is a very interesting question that no one knows even a fraction of the correct answer to. What I am sure of is that as and when this happens, the entire economics of the world will evolve, and humans will surely find something more productive and interesting to spend their time, effort, and energy on.

If we change our thinking from what robots can do to the place that robots will have in our society, we are on a very different line of thinking. Can I make my humanoid my best friend, fall in love with one, or get married to one? It all feels weird and strange – so, what will make these humanoids acceptable in our society? What will it take for humanoids to win our hearts and once they do, will they coexist with us? I will borrow an observation from the human world. We like to associate with and meet people who understand us, don't judge us, listen to us with empathy, and give us what we want. After all, that's what makes an experience delightful. The following figure presents the varied levels of expectation that humanoids will inherit from humans:

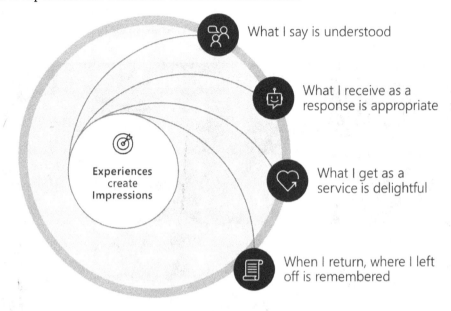

Figure 15.6 – Humanoid experience chain

I will summarize the four key elements that should be embraced by humanoids to be successful in the world of humans. This is over and above the ground rules of engagement. The humanoid should understand me and my needs, listen to me, and not operate in a question-or-answer format. The response I get from a humanoid should be appropriate, complete, and meaningful to my context – a situation in which humans, humanoids, and digital beings coexist is not so distant. I also expect a humanoid to remember me and pick up from where we left off like a human would, to apply what it has learned, and greet me with a smile. That's what will be the start of a human and humanoid bond. Yes, a bond with the machine exists since it remembers me and knows my preference. Or is this just my imagination? Only time will tell, but surely 2050 will be very different from what we can see right now.

Summary

I think by now you must realize that what I am sharing with you here is no longer science fiction. You may also be thinking that if digital humans, humanoids, digital avatars – which, in the metaverse, can be called a more intelligent you – chatbots, and converged networks with 5G and 6G come into play and mix with physical reality, it'll become next to impossible to differentiate between what's real and what's virtual. While professors and inventors across the globe are busy researching and writing algorithms to inject emotions into machines, only time will tell how this will evolve over the coming years. It's currently uncertain whether the impact work beings will have on society will be similar to the impact social networks have had on society. The effects are yet to be determined. In any case, the goal is for machines to interact with humans in a friendly and engaging way, creating memorable experiences for both human beings and work beings.

In the next chapter, we'll explore further how all of this will come to life, the risks it may present to us as a society, and the role that ethical AI and responsible programming play.

16
The Role of AI in Managing Future Lockdowns

Let's visualize what the world will look like 10, 15, or 20 years from today. According to the United Nations, 2 decades from now, we will have around 9.7 billion humans and over 25 billion internet-connected devices. That's 2.5 connected devices per human.

Today, what comes to our mind when we talk about devices are phones, laptops, smartwatches, and digital assistants, such as Amazon Alexa or Google home devices that are constantly connected to the internet. All phones today work in the same way and look the same. Screen touch is the primary input interface and the display is the primary output interface. With **artificial intelligence (AI)** becoming mainstream, digital devices' interfaces are bound to change. AI will create more devices and more form factors, enable device-to-device communication, and introduce new input and output interfaces for interaction with machines.

In the past 2 decades, laptops have not seen any change in their form factor other than increasing memory or processing power. Laptops today still look the same, with a keyboard and a mouse or a touchpad as the primary input devices. You can say a tablet is a new device, but it is essentially still a big-screen phone with touch as the primary input. The point is that AI will transform tablets and laptops by transforming the input and output interfaces. You will start seeing robots in your neighborhood more frequently, which will provide a new form of computing. A new category of transportation computers will arise in the form of connected self-driving cars, bikes, and utility service vehicles. Most of us will talk or just share our expressions to act as input to these new forms of devices for them to do their job.

These new devices will be able to do more than just display; possibly, they will talk and do physical maneuvers as well. As voice becomes a primary input interface, we will struggle to validate whether we are talking to a human or a machine. Machines will use software that is currently categorized as *deepfake* to reproduce voice and video, giving interactions a sense of familiarity. Deepfake borrows its name from two terms – *deep learning*, which is a subset of AI, and *fake*, which means fraudulent or a hoax.

Humans' interaction with machines and software across various hardware formats is bound to impact society and different relationships. In the last 10 years, we have seen billions of users across countries start to use social networks and touchscreen-based phones. The use of social networks has already created new habits, such as checking in at locations we visit for business or pleasure, sharing images with our friends and others on social networks, or spending a good part of our time surfing the internet or apps using the touchscreen on our phones. Innovation across hardware and software spaces will pose new kinds of risks for human beings. As we are starting to recover from the lockdown induced by the COVID-19 pandemic, we must also prepare ourselves for lockdowns induced by AI in the future. These will be lockdowns triggered by AI machines that have the power to pause the internet, disrupt the availability of social networks, make our touchscreen nonfunctional, disrupt electricity, randomize traffic, change the target destination of self-driving cars, or trigger a lockdown in a town, city, state, or country. Such lockdowns may come with new terminology; I call them *digital lockdowns*, and they pose different kinds of both physical and mental risks for us. What we need in response are laws and regulations that can prevent, mitigate, and lay down ethical practices, which should be followed by AI software and technology companies that operate such technologies on a global scale.

In this chapter, we will dive into the world of the future by covering the following topics:

- Input and output devices
- Digital assistants – at home, for travel, and at work
- Family time and social life
- Education and study time
- Healthcare
- What should humans do?
- The role of AI and ethics

Input and output devices of the future

I believe AI will play a key role in introducing new input and output devices in the future, and changing those that already exist. Today, input devices, primarily a mouse and a keyboard, are mechanical. I think computers in the future will constantly be looking at us, reading our expressions, observing our hand and facial movements, and interpreting them like humans. This will be dependent on multiple factors, such as our daily habits, prior usage of computing devices, and whether we work in a home or office to make an as accurate as possible prediction. So, what that means is that in the future, devices may be able to observe human behavior, read facial expressions and hand movements, and interpret expected actions that humans want devices or other computers to perform.

Computers also interpret commands and instructions from humans, taking into consideration the external environment, such as businesses, the economy, the weather, and what's happening currently in a city or locality that you reside in.

Most of us start our day by checking our email and selecting the priority emails to reply to or act on. However, why do we need to read emails in a linear fashion and then respond? The email system of the future, based on your behavior and email content, will show you emails that need your attention filtered by priority. The email systems of the future will also read emails that require an urgent response, get necessary data from your email inbox and other systems, recommend decisions for you to take, or reply on your behalf. The email system of the future will also operate with a voice and talk to you, making suggestions and taking orders for replies.

With just a few clicks of the mouse, we can easily switch from email software on our screens to browser software and search for information instantly. We then end up searching, clicking, and reading in the browser. This kind of information search and response needs lots of human time and attention. This occurs because our primary interface is a mouse, a screen, and a keyboard. The browsing of the future will be different. Today, it's time-consuming to even go beyond two or three Google pages when we search, and we tend to get lost in an ocean of information. Browsing in the future will involve a conversation with a user on which parameters to refine. This means asking about our intent – for example, if I am searching for a car, the computer will ask whether I am looking for a new or used car, what my budget is, and whether I need to buy one now. The browsing system of the future will have access to my financial situation, which includes my bank balance, my credit balance, and my credit score. It will either estimate or ask what my budget is as part of the search. The browser of the future will also be aware of the color of the car that I prefer, either based on my browsing history, posts on social media I have liked, or conversations that I have had with anyone in past, and come up with accurate recommendations. It will use all this data to search and then bring up a list of cars that I can buy that are close to my current location. How we search for information has not changed in the last 2 decades, and I am confident the new types of hardware devices will also come with new ways to search. You will not spend time clicking a mouse and opening various browser screens. The browser will come back and converse with you like a human through the voice feature, chat, or user interface, where you select your choices for the browser to display precise results. As you listen to the search results, the browser will be able to read your facial gestures to show you more or move to the next precise search option. You won't have to click the mouse to go to the next option.

Now that facial gestures are becoming popular as one of the input interfaces for computers, it will have an impact on how we authenticate to a system. While facial gestures today on mobile phones might be primarily used to open the lock screen, in the future, they will go beyond just logging into a system. They will use the surrounding environment and additional context to predict what a user expects, such as music. The gesture system will recognize your surrounding, such as the gym, and will automatically play your gym music. It will recognize you are in a meeting, and mute non-critical notifications and reminders. The impact of gesture recognition will be different across various form factors, from a phone or digital assistant to a self-driving car or digital lock on a front door that opens automatically, based on your retina scan and gestures. When a computer can see, hear, and observe contexts, such as the time of day and the weather, and has access to what most of us prefer in those contexts, it opens up new options for computers to engage and interact with humans, which would depend upon not just what the users see (a screen) but also what they touch, type and click. This will introduce comfort and convenience, but also a new type of digital risk.

Digital assistants – at home

For decades, the alarm clock was responsible for waking up millions of people before we started relying on mobile phones. Today, a voice command does the trick. A voice command to your phone or home digital assistant not only wakes you up with an alarm but also provides you with the daily news or a weather report, as per your request.

Fast-forward a few years and I'm sure digital assistants will be running our homes, using mostly voice and voice recognition systems as the primary input interface. The digital assistant will know everything about our homes, such as the number of rooms, heating and cooling systems, lighting and preferred lighting colors, cameras for internal and external surveillance, and kitchen-connected devices. Additionally, digital assistants of the future will be able to recognize family members, pets, and other robots authorized to live in the same home. The digital assistant will also have access to our office and personal calendars, helping us to be productive both at work and at private appointments throughout the day. Based on your daily patterns, these assistants will be able to predict your needs and operate accordingly.

In the good old days, when I got back home and looked tired, my mother would offer to make my favorite dish to cheer me up. Now, if my home digital assistant notices that I am not happy, it will send a signal to my personal robot. The robot will try to initiate a conversation with me, asking how my day was. The home assistant or home robot will scan my gestures and use a predictive model to guess that I may want to relax. The robot will try to search what I did in the past when I had the same expression and will respond accordingly. For example, if I listened to jazz music in the past, it would ask me if the digital assistant should play the music. If I chose to take a short nap or order takeaway food, it will ask my permission to do this again. It's like reading my mind.

Today, it's very difficult for me to confidently say that this relationship between humans and machines will be universally loved by billions of people across the globe. However, I am very confident that the generation born after the pandemic will have experienced computing and AI as an integral part of their life growing up. They will surely love and appreciate these gestures from robots and other machines. It's the same preference gap that exists today in the choices we make about phones, cars, and gadgets compared to our parents.

I have no doubt that home digital assistants in the future will be capable of doing more than just switching on lights and playing music, and will offer a better and more comfortable life to humans. What does intrigue me is the risks that it will bring in terms of machines knowing so much. This includes the risk of an attacker knowing our moods, preferences, and actions, and the risk of machines suggesting actions that have malicious intent.

My digital assistants – for travel

The rise in digitization will not just be limited to our homes. It will also change the way we travel. Traveling involves transportation, and although people will still own cars in the future, travel is set to change with pods. A *pod* will be a driverless, autonomous car with digital cockpits. Pods will have luxury seating, large touchscreens, ambient lighting, and economy and business seats, offering you privacy and very high-speed internet connectivity besides transporting you from one point to another. Pods will mostly operate on a subscription model, allowing people to enroll in transport services and facilities. They will be able to travel on roads and in high-speed underground tunnels. They will run based on dynamic frequency, as they would be able to calculate how many people are at work and offer a predicted time for when they return home. Pods will also operate inter-city, and they will be able to use legacy railroad systems to travel between cities on high-speed train networks. One day, pods may be able to fly, but I doubt that is likely in the next 15–20 years on a global scale.

Pods that operate with a driver will introduce a different kind of risk. It will be interesting to watch what happens when a driverless, autonomous car commits a traffic violation. Who will pay the fine – the pod operator, the pod manufacturer, or the subscriber who asked the pod to speed up, as they were running late for a meeting? If we look at scenarios where a car ends up in an accident due to bad road conditions or after trying to overtake a manual-driving car, who ends up being responsible? What will be the manufacturer's liability if the operating system of the car that is responsible for autonomous driving malfunctions results in damage to a property or another car, or loss of life?

Sitting in a vehicle that gives you the privacy of your own car with the comfort of business-class air travel is an inevitable luxury of the future, but it brings with it a new set of regulations. I believe the transport sector at large, starting with car manufacturers, will have to go through lots of regulations and certifications as they move toward the adoption of autonomous digital vehicles. Personally, in the future, I would book my pod to the airport if it became available in my city. You can learn more about such vehicles here: `https://techcrunch.com/2022/04/20/elon-musk-mass-produce-robotaxi-by-2024/`.

Autonomous vehicles, also known as pods, have been introduced gradually in waves, gaining momentum over time. During the early stages of pod deployment, human drivers may accompany them as backups and take over if necessary. However, within a year or two, larger pods or autonomous trucks may travel down interstates without any human drivers as backups. It is becoming increasingly apparent that the driverless revolution will progress slowly, mile by mile, and neighborhood by neighborhood. For most people, riding in an autonomous vehicle will be a new experience, potentially beginning with placing trust in specific roads or highways at certain times of day and only in favorable traffic and weather conditions.

A TechCrunch report on Elon Musk's vision of pods has increased my confidence in riding in an autonomous vehicle in the next 5 to 10 years.

Digital assistants – at work

Today, we can look at our calendars and then log in to our virtual meetings. We click a few buttons and enter a meeting ID or password to log in to a virtual meeting, using applications such as Microsoft Teams, Zoom, and Google Meet. The meeting applications of the future will work with the help of gesture recognition. What this means is that I will not have to click the mouse button to log in. The meeting software login module will use my camera and read my facial expression or gesture to log me into the virtual meeting. My face is my password to authenticate, so I won't need to type any password. Based on my hand gesture or voice, the application will choose to show the real me or my avatar in the meeting.

Today, when we participate in virtual meetings, we may not be physically present in the meeting, but we are mentally connected and fully immersed for the duration of the meeting. If the meeting is a video meeting, then you need to stay within the view of the camera to ensure your presence in the virtual meeting. However, the future of virtual meetings will be different; you will be able to send your virtual avatar to represent you in meetings instead of logging in yourself. Your virtual avatar, also known as a mini-me, will attend the meeting on your behalf, take notes, stay connected with you, and even bring you into the meeting as required. Your virtual avatar will have knowledge about you, your priorities, your ways of responding, and your past decisions. You may also authorize your virtual avatar to take decisions on your behalf, if they follow a pattern. If you choose to send your avatar to a meeting and it needs some help, clarification, or a decision, it will converse with you privately to get your approval on what it communicates on your behalf in the meeting. You will be bound by your avatar's decision, as it's trained by you and mirrors your imagination, thoughts, and decisions. Almost all computing devices in the future are likely to have input-output devices that can recognize voice, eye, and hand gestures, gestures from digital assistants, and environmental context. You will still be able to attach a keyboard and mouse to a computer, but voice and body gestures will do most of the interactions. Almost all email replies, communication, and meeting notes will be created, followed up, and communicated by a meeting software digital assistant. Merely by using a glance, I will be able to train a system. The system will learn from changes in my voice modulations as I speak, and I can check whether the machine is interpreting and learning the emotions of my facial expressions. Digital assistants will not be limited to note-taking and sharing after meetings. Digital assistants of the future will also track and arrange resources for items to be actioned, predict which meeting participants will need to complete action items, and regularly remind them to get them completed in time. It's like having the world's best personal secretary that ensures we don't procrastinate and complete all the work on time.

Just like we learned to use *touch* to operate mobile phones a few years back, we will be using body gestures in the future across computing devices. I am confident that the majority of us will quickly adapt to using touchscreens, effortlessly swiping left and right using new user interface touchscreen control options, as well as using hand gestures and other physical expressions to interact with future computing devices.

As we attend virtual meetings today, it's very common for someone to speak while a computing device mic is muted. What comes next is someone calling out, *"You're on mute!"* Most of us can empathize with and relate to this prompt. On future computing devices, such nudges will be history. If you are talking about something that is relevant to the subject at hand, you will be automatically unmuted, even if you had muted yourself. You will feel as if some superhuman is controlling the meeting, and the device will know which voice to broadcast and who to mute, such as a person in a noisy environment adding to the background noise of the virtual meeting. We will feel as if digital assistant software can see and listen to all participants, regardless of their mute or video status, controlling and creating a better meeting experience for everyone. If there is background noise, this superhuman digital assistant will suppress it; if your video is off and you are not in your seat, it brings up your avatar. Moreover, if you are discussing a sensitive topic in a virtual meeting that shouldn't be heard aloud, this superhuman digital assistant will request all participants to wear their headsets for privacy and security. This is the ambient technology of the future in action.

I am confident that our screen-sharing experience will also change in the future. Another common phrase in today's virtual meeting is *"Are you able to see my screen?"* Most of us when in a virtual meeting either forget to share the screen or share the wrong screen, particularly if we are using multiple screen monitors when working on a computer. Today, we have technology that can use AI to accurately detect who is talking. This will also help to build new software intelligence that will automatically share your screen, as the software listens to a human voice that says, *"Let me share my screen."* So, instead of you manually taking a few steps to share your screen, the software, once it is ready to share your screen, can just ask you for consent and share it with all other attendees automatically.

In virtual meetings in the future, automation will play a significant role in a digital assistant's capabilities. It will be able to share your screen automatically, select the appropriate screen if you are using multiple monitors, and ensure that it is visible to all meeting participants without the need for manual intervention. Meetings in the future will be way more valuable, productive, and convenient than what we have today. They will be a hybrid model, and participants will consist of humans, bots, avatars, and digital assistants of attendees. Bots will, in real time, align themselves with ongoing discussions based on the meeting context and determine post-meeting actions to take on behalf of the department or human they represent. It will be a different kind of experience compared to what Generation X has grown up with.

Virtual meeting software will have meeting analytics that calculates your meeting efficiency, talk time, decision time, analytics time, response time, and emotional balance using your facial gesture. These statistics will mark and map your growth and improvement areas quarterly and yearly, giving you real-time guidance and suggestions. It will be, after all, a data-driven world.

All this digitization can also pose several risks. For example, sometimes, a user joining a meeting might be a deepfake impersonator instead of the real user or their avatar. Let's suppose you're getting a video call from your dad, and he informs you about his deteriorating health and asks for monetary help to cover the medical bills or other expenses. What if it turns out that the video was a deepfake and the sender was intending to extort money from you? I am very confident that independent verification companies will spring up to create, monitor and endorsing content and verify a digital avatar or real human in real time as they communicate. Such verification and validation services will create a semblance of sanity by creating a sensible form of regulation in contrast to today's unregulated social media markets, and it will also diminish the abuse of social media in the name of freedom of speech. The abuse of social media and freedom of speech is linked to the proliferation of deepfakes. Deepfakes are manipulated videos or images that can be used to spread misinformation or propaganda. They can be created to deceive people or damage someone's reputation. Social media platforms are struggling to combat the spread of deepfakes due to the ease of access to technology and the difficulty of detection. The misuse of freedom of speech to spread deepfakes can have serious consequences, including political instability and social unrest. As such, it is important to raise awareness about the dangers of deepfakes and develop effective strategies to combat them.

Family time and social life

The metaverse or the digital virtual world, with its avatars, digital shops, digital roads, digital or virtual cars, and so on, is still in its infancy at the time of writing (October 2022). Fast-forward a few years ahead, and kids playing and learning in the metaverse will be the norm. Even today, my 8-year-old daughter spends a lot of time on Roblox, which is an avatar-based game. Shopping in the metaverse will be the new normal, where you can dress up and see how you look when you are at a virtual airport, a virtual cafe, or walking down a virtual street from various video vantage points. It will give you a complete visualization before you make the actual purchase of that dress, suit, bag, or cycle in real time. In years to come, you may not go to shopping malls as frequently, as virtual engagement becomes the primary interaction between brands and consumers. Today, we call this *augmented reality* or *mixed reality*. I don't know what the naming taxonomy will be in the next few years, but what I sense is that 3D projection technologies will become built into our phones, watches, mirrors, and almost all the digital screens around us.

Large brands will let you experience 3D visualization using new types of screens and projection technologies, where you will not need to wear any HoloLens or similar devices. It will be difficult for human eyes to distinguish between what is real in these new formats of digital virtual stores. Sensors in a store will monitor each movement of your body, including hand, eye, and voice gestures, to create a compelling real-life engagement in the customer purchase journey. Going to a branded store in a mall as a family will be like going to Disneyland to watch your favorite character show in 3D or 4D. Window shopping will reduce, to be replaced by extensive mixed or extended reality-based window shopping that will be done remotely. You will request a store to show you a specific dress, for example, and you will be able to try it out and visualize how it will look on you in seconds, based on your body posture and shape. This is only possible in the real world if you go to the store and try it out for yourself. The real world will recede in place of this extended world, and kids born and brought up in this era will wonder how our parents used to shop 3 decades back, much like how we today wonder how our grandparents lived without mobile phones and Google Maps.

In most countries today, you can easily spend months without making a physical cash transaction. In a few years, money might go completely digital. From purchasing groceries, traveling in taxis, and paying rent to home loan payments, purchasing travel tickets, and other utility transactions, you will rarely transact using cash. As well as cash, payments today are handled via credit or debit card. As society becomes more digital, your phone will become your credit or debit card. In years to come, you will become your credit or debit card, using your unique retina or a facial scan. Payment machines that today have a slot to read or swipe cards will instead have a camera to look at you, show the amount that is requested for payment, and based on your gesture complete the transaction. Your biometrics will become your passport, credit card, your bank account, and your share trading account. Everything will be tied to your biometrics, removing the need to carry any plastic cards or government-issued documents such as a driving license. However, I think people will still carry documents or plastic cards, just in case the digital systems don't work as expected. This is the risk when you go completely digital – what if a system does not work as expected? What if a payment system is unable to recognize you to authorize a payment using face gestures? You may fall back on using plastic cards or PIN-based authorization in such cases.

This digitalization and AI will spread to every point-of-sale system, including grocery stores, autonomous taxis, ATMs, and banks, all of which will be able to identify you in real time. Every system will be connected. This comes with lots of benefits and lots of risks. Connectivity disruptions are rare, but when they happen, they can bring a whole city to a screeching halt.

We will also have to mitigate and be ready with a risk strategy in place when banking and financial systems are attacked, networks are hacked, or verification and authorization systems develop bugs or experience planned or unplanned downtime.

Education and study time

The learning and education sector underwent massive change during the COVID-19 pandemic. Education in schools, colleges, or on any professional course for centuries has been associated with a face-to-face, in-person experience. The pandemic forced a change in learning and education to virtual learning experiences. As we look into the future, the pandemic will become an inflection point for this sector. It's easy to visualize learning in the future coming from any location, using more immersive metaverse or 3D experiences.

In the future, students will have the ability to learn from multiple teachers – some real, and some virtual. Each student will be on their own learning journey, learning at their own pace. The virtual teachers will ask questions in real time and explain topics based on a student's response. With advancements in virtual reality technology, students can now experience learning in a whole new way. Instead of just reading about a concept or watching a video, they can actually immerse themselves in a virtual environment and see how it applies in real life. For example, a history lesson on ancient Rome can be brought to life through a virtual tour of the Colosseum and the Roman Forum. A science class can take students on a virtual field trip to explore the inside of a cell or the depths of the ocean. By experiencing these concepts firsthand in a virtual setting, students are more likely to retain the information and develop a deeper understanding of the material. Additionally, virtual reality can provide a safe and controlled environment for students to practice skills such as public speaking or conflict resolution. Overall, virtual reality has the potential to revolutionize the way we learn and provide students with a more engaging and immersive educational experience. This may change the need for students to travel to college, school, or university every day. Like work, education will also transform in that most of the work will be done remotely or in a hybrid fashion.

Learning systems to create more engagement will deploy gaming and AI engines that will reward the learner after each interaction. It will remember learners' preferences and expressions to make it more interesting for them by customizing the learning process. Learning systems will use this information to create a learning path for each student. Teachers will be able to clearly see the time spent by each student and the methods of teaching more suitable for different kinds of learning personas. Like all digital systems, education will also be riddled with its own risks. There is a risk of whether a learner is really learning or just browsing the internet. There is also a risk related to verifying whether the real learner is the correct human being who enrolled for a course or someone else, a risk related to humans transitioning from in-person to virtual-only or hybrid virtual learning, and a risk related to a master learning system content becoming biased or containing poisonous content that results in an unpleasant learning experience. Similar to how social networks today are able to sway our thinking, learning systems will run the risk of taking a singular view toward learning topics. Finally, there is also a risk related to when digital learning systems come under attack, resulting in the closure of schools or colleges across a city, state, or country.

Healthcare

Another sector that got directly impacted during the pandemic and that is transforming itself faster than others is the healthcare sector. During the pandemic, remote health consultation became a way of life due to government-mandated restrictions on humans going outside their homes. Digitization, social distancing, and travel curbs led to the birth and rapid advancement of remote monitoring devices. Devices came in many forms, from wearable and stickable to injectable.

In the future, wearable devices will come in the form of watches or bands that have the ability to capture vital statistics and send data via connected networks to patients. Patients will then use this data from the devices to receive recommendations from a doctor, virtually or in real time.

We will also have stickable devices that can be used once and thrown away. Simply put, they are disposable. These will be lightweight, sticker-type devices. They will stick to your body part and collect vitals such as your ECG, body temperature, pulse, breathing synchronicity, and other critical parameters for further diagnoses and recommendations.

Our home digital assistant will have add-on devices to provide statistics about our health, with technologies such as sonar, thermal, photon, and x-ray that can determine body temperature, fractured bones in 3D, blood loss and stream rates, and other vital parameters for further diagnoses and consultation.

In the field of medical devices, there has been a significant advancement in recent years with the development of high-tech devices that can analyze human body functions with incredible accuracy. For example, there are now devices that can measure a person's blood pressure, heart rate, and oxygen levels using non-invasive methods, making it easier for medical professionals to monitor their patients' health. In addition, devices that can analyze a person's eye pupils, tongue color, texture, or formation are becoming more prevalent, allowing doctors and healthcare providers to detect potential health issues earlier and with greater accuracy. These devices are especially useful in remote or underserved areas where access to medical professionals may be limited, as they can provide vital health information quickly and accurately. With the continued development of these devices, medical professionals will have more tools at their disposal to diagnose and treat patients, leading to better health outcomes for everyone. The use of these devices will eliminate the need to visit a doctor and make remote medical consultations possible by sharing their output.

These devices will have sensors to check for temperature and humidity. Cameras in these devices will help track people's movement, sleep, walking, and cough or cold patterns to give parameters if required for virtual or actual doctor consultations. Devices in the future will have precise monitoring of what we ate, how much we smiled, how many times we sneezed or coughed in our house, and our energy and sleep patterns in detail. It's like being watched 24x7 by machines that, in seconds, can share critical information with a doctor to give you the right advice.

One set of devices that is worth mentioning is the type that can collect blood, urine, or saliva samples for detailed diagnosis. I am more confident that in the future, with more technical advances, there will be a reduced need to collect blood samples for analysis. Bosch has built a hemoglobin monitor that does not require any blood samples from patients and provides accurate hemoglobin levels in your blood. (You can read more about this here: `https://www.bosch-presse.de/pressportal/de/en/bosch-hemoglobin-monitor-early-detection-of-anemia-without-blood-tests-223296.html`.)

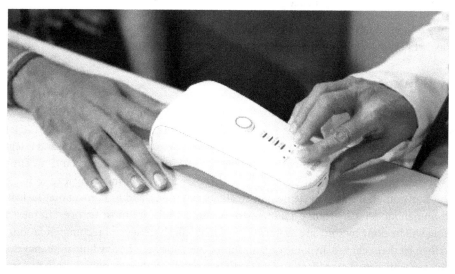

Figure 16.1 – A non-invasive hemoglobin monitoring device

In the future, I also believe we will not have as many hospitals. We might be moving toward a future where, as a society, we do not visit large unfriendly buildings called hospitals as often. I am not saying that we will not need hospitals. We will, but we will require fewer of them, as the majority of doctorly advice and medical services will be served by digital humans, bots, and special-purpose medical robots and machines that can be purchased for home use.

Technology-infused AI services will become active participants in our wellness and health decisions from the comfort of our homes. I also believe that we will not visit a doctor for a physical examination, due to high-precision wearables or new forms of devices that will be commonly available at the local chemist or drugstore for home use. Almost all homes and communities will have basic-to-moderate hospital facilities, remotely operated by teams of qualified doctors and AI virtual doctors as a monthly or annual subscription service.

Almost all innovation in this space will rest on high-speed mobile internet connectivity, new medical devices, the use of data science predictive models, and human expertise to give society access to the best doctors across the globe, anytime and anywhere. As with the other subjects covered in this chapter, including remote working, travel, and education, these innovations in healthcare will give birth to a new set of risks that may have an impact on society. These risks will include compromised devices, leaked sensitive medical data getting into the hands of attackers or entities that can profit by using it, and patients receiving incorrect medical advice due to the tampering of data by an attacker. Would it be wise to completely shift to an AI virtual doctor if it brings the risk of it being compromised and acting on the instructions of an attacker? The governments, device manufacturers, and medical services providers that undergo these digital transformations will need to analyze the new risks that come with these new devices and capabilities, finding ways to mitigate them to provide safe medical advice and patient management practices.

It is certainly interesting to think about what humans will do with so much digital innovation and automation. Let's explore this in the next section.

What will humans do?

Digital life will augment human capacity and disrupt or depreciate old ways of living, studying, exercising, traveling, and working. In the years ahead, most work will be automated, and repetitive work will be done by machines. As machines and automation become more advanced and integrated into our daily lives, there is a possibility that we may become increasingly detached from our work and the processes involved. While the efficiency and accuracy of machines may lead to increased productivity and output, the human element of creativity, emotions, and fulfillment may be lost. The human-to-human model will become a human-to-machine model for most services. In sales, most of the selling will be done by software avatars or bots that will share several recommendations with a user and then let them choose. In some organizations, there may not be any human manager. They will get replaced with virtual managers, and most of them will be AI-based software with the promise to improve an organization's efficiency, revenue, and output.

For most organizations, technology will become a critical element to ensure efficiency in operations by using AI and predictive models. Most schools will not employ human teachers and will find virtual teachers much more suitable when it comes to teaching and being able to provide a learning experience to students at any time. A virtual teacher will treat each learner based on their ability to learn, their preferences, and their language, and will be available 24x7. The virtual teacher will be much more illustrative and engaging than a human teacher.

Agriculture and farming in almost all countries will be made better and cheaper by agriculture robots. Humans won't have to work in factories, as everything will be done by robots. Almost all taxis and trucks globally will be autonomous and will not require humans to drive them.

Most experts predict that digitization and AI will amplify human effectiveness but might also threaten human autonomy, psychology, relationships, and social and work fabrics. In previous chapters, we have seen how machines can potentially exceed human intelligence and capabilities in many tasks, including complex decision-making, visual or speech recognition, language translation, computation, pattern recognition, and even making sound decisions as a virtual doctor or lawyer.

Cars that drive themselves, machines that read medical reports, X-rays, and software algorithms that respond to customer-service inquiries are all manifestations of powerful new forms of AI and automation. Will this make humans less needed for certain jobs? I don't know the definitive answer, but I will try to share my thoughts, assuming that almost all repetitive work will be done by machines.

Will digitization and automation reduce jobs and increase inequality, or will they bring more significant work and economically well-off societies? This question has worried humankind ever since computers and machines took over certain manual jobs 3 decades back. The services of a file keeper, travel desk operator, switchboard operator, and many more are no longer required, as we have computers, software, and mobile apps doing these jobs faster and better. The question still remains, what will the future of work look like and what will humans invest time in?

I believe the economic pyramid will change as it has changed in the past. If we go back a few centuries, we had most people going to factories and doing manual jobs. Most people who worked in factories used to commute from nearby towns, by either walking or cycling. There were no cars, and most long-distance commutes were done on carts powered by horses or cattle. Then, the motor car was invented. This disrupted the need to use animals for long-distance commutes.

Car factories arose and generated jobs. New modes of transportation arose as a result of cars, such as buses, trucks, airplanes, and other modes of transportation. This generated jobs, uncovered needs to travel, facilitated travel across cities, and generated jobs. More jobs generated more engagement for the population at large. More humans got engaged and found passion in what they loved to do and the living they earned.

Mobile phones have been in our lives for the last 2 decades and have transformed the way we communicate with each other. Mobile phones created new digital stores, new phone carrier companies, and the new economics of mobile phones and associated devices, mobile phone apps, and a large app-developer ecosystem. This ecosystem of app developers created both IT and non-IT jobs.

On the IT side, it created multiple new jobs, including mobile phone application graphic designers, testers, mobile application usage analysts, and mobile application marketing business roles. The mobile ecosystem on the non-IT side transformed tasks such as food delivery, taxi booking, grocery ordering, and purchasing last-minute hotel rooms or ticket offers. This new form of engagement generated more economic activity in other non-IT sectors of the economy. Food delivery businesses generate more jobs and opportunities for food delivery personnel, with more delivery vehicles and more choices for customers to order from. In short, the mobile revolution did replace landlines, but it also dramatically added new jobs and increased the overall remote talk time between humans. I strongly believe the same story will play out in the coming decades, providing clear answers on what humans will do, as digitization, automation, and AI change the way we work, travel, and live.

I can comfortably predict that we will have new job roles that don't exist today and that we are unable to imagine. These new job roles in the coming decades may have a new economic model and impact industries the way the mobile has in the last 2 decades.

Let me share some skills I believe will come into existence in the coming decades.

In the future, there will be a wide range of roles and positions available to those interested in creating humanoid robots. Some individuals will specialize in creating the realistic skins, eyes, and body parts necessary to give robots a more human-like appearance and feel. Others will focus on machine psychology and developing emotional responses that will enhance the overall experience of interacting with robots.

Students will learn how to apply principles from physics and mathematical modeling to develop drones that can fly and carry heavier loads in varying temperatures and air pressures. In addition, some professionals will spend time working with doctors and medical practitioners to integrate technology implants with the human body and other living organisms. One potential application of this technology could be the creation of memory implants to help treat conditions such as Alzheimer's, which affects many elderly individuals.

Advancements in nanotechnology and implants will create new job opportunities in fields such as biotechnology, gene therapy, and DNA biosensors. The devices needed in these fields will be powered by the energy generated by the human body and will require the development of new types of chips, sensors, and other compatible devices. However, as more tasks are automated, there may be less room for human-like emotions, smiling and joy, or fulfillment in completing work tasks alongside machines.

Yes, I do believe that as most jobs will be taken by machines and AI, our aspirations will move toward something bigger, such as building transport systems that can take us to other planets in search of civilization. This will generate new types of jobs, new roles, and new skills because human beings won't be consumed with trying to fulfill basic needs and requirements.

In such a scenario, our large working population will need an open mindset to leave their current jobs, quickly gain new skills, and then migrate to new roles and jobs.

Personally, I am less worried about job loss or what humans will do. I can say with confidence that we will have more jobs and roles in the future, but at the same time, some roles and jobs will become redundant. What I fear is the potential risk that such widespread digitation, automation, and use of AI will bring to our society. As we recover from the pandemic-induced lockdown, our society, government, and technologies need to prepare for a response to a lockdown forced on us by machines and AI software. Let's look at the impact this could have on society in the next section.

What is digital shutdown?

The COVID-19 pandemic forced governments to apply desperate measures, namely lockdowns. These resulted in closing malls and retail outlets, schools and universities, government, and corporate offices, taking transportation off the roads, and ordering people to stay at home. This was done to avoid human-to-human interaction to prevent the spread of virus infections.

Staying at home has social impacts. Businesses such as restaurants don't get customers, thus forcing them to close. Retail outlets don't sell any clothes or sports goods items, or even cosmetics and personal care items such as perfumes, resulting in low consumption and impacting manufacturing. Daily workers are unable to get jobs, resulting in a living crisis, including a food crisis for poorer people.

I still recall the South by Southwest tech conference in Texas in March 2018, where Tesla and SpaceX founder, Elon Musk, issued a friendly warning: *"Mark my words, AI is far more dangerous than nukes."* How, exactly, could AI become dangerous? The rationale is that the smarter machines become, the more autonomous they become. As you become more autonomous, you can decide to change your goals. What if the changed goals are not aligned with human peace and co-existence? This is the point at which computers, machines, and software reprogram themselves and successively update their goals, leading to a so-called "technological singularity" or "intelligence explosion." Once machines operate at this level, the risk of them outwitting humans in battles for resources and self-preservation creates a scenario that's not easy for humans to mitigate.

Let's say in the future you buy a robot to clean your house. The robot observes that you as a human are the reason for the house being dirty. Your eating habits result in trash that needs to be cleaned. You keep moving in and out of the house, bringing in dust that needs to be broomed by the robot. The robot has a singular goal to keep the house clean. It decides to lock you in the basement forever. The robot is now working toward its goal, but it's placed you in danger. What if your city was run by a robot AKA AI software? Water, electricity, traffic lights, transportation, farming, and education were all operated by AI software controlled by the government. One day, this AI software one day decides that in the interest of preserving nature and natural resources, it's implementing a shutdown. This time, it's not a pandemic-induced shutdown; it's an AI software-induced, city-wide shutdown. I call this a *digital shutdown.*

An article on Vox.com (https://www.vox.com/future-perfect/2018/12/21/18126576/ai-artificial-intelligence-machine-learning-safety-alignment) starts with a Stephen Hawking quote that says, *"The development of full artificial intelligence could spell the end of the human race."* This article imagines an extreme case of a digital shutdown, with the development of a sophisticated AI system that realizes it can do more if it uses all the world's computing hardware, and it realizes that releasing a biological superweapon to wipe out humanity would allow it free use of all the hardware. That's more dangerous than a government lockdown or my cleaning robot locking me in the basement.

The idea that AI can become a danger is rooted in the fact that AI systems pursue their goals, whether those goals are what we really intended — and whether or not humans are in the way of achieving those goals.

It's rare that I encounter AI developers, technical wizards, and visionaries in AI who think a digital shutdown is impossible. Instead, I get a message from them that it will happen someday — but probably a day that's far away. The answer to whether or not a digital shutdown is possible lies in how we are building AI today. The answers lie in the ability to create thinking machines that are aware of ethical issues and dilemmas, ensuring that such machines do not harm humans and other morally relevant beings, and giving the machines a set of morals. Yes, we are talking about machine consciousness and the role of ethics in AI.

The role of ethics in AI

Is it fair for a taxi operator such as Uber, Lyft, Ola, Grab, or so on to charge you surge pricing? Let me rephrase the question with technical jargon – is it right to make a software algorithm that shows and charges you a high price when it knows that there is no other taxi nearby to take you to your destination? Here, it is critical that the software knows that its taxi operator has a monopoly on your custom and, hence, can decide on any price to charge you and maximize profits. Should the algorithm be made so independent that the taxi can charge 10–20 times the normal fare, or should we have regulations that limit the surge price to a maximum of 2–3 times the normal fare? We are now entering the world of ethical responsibility in software, AI, devices, hardware, and practically everything that falls under digitalization and automation.

Here is what I think should be done by governments, communities, and users, in general, to ensure AI is used ethically and in a productive manner in technology:

- **Regulations and control around AI and supply chain automation**: When humans make a mistake, there's usually an inquiry and an assignment of responsibility, which may impose legal penalties, termination, or suspension of the human decision-maker. This helps the company or community understand what's right and build trust with its stakeholders. In the future, I believe we will have regulations that will enforce AI owners and operators to explain their decisions. As the use of AI becomes more widespread, it will be necessary to address the issue of machine biases that can occur in these systems. To ensure that AI systems are fair and unbiased, sharing the source code in public or during litigation may become necessary. This will allow experts to identify any biases in the training data used to teach the AI system and make the necessary corrections to ensure fairness.

  Moreover, transparency in the data used to train AI models will become increasingly important. As AI systems rely on training data, the data that is fed to them and its impact will also come under the purview of disclosure. This means that companies and organizations that use AI will need to disclose the sources of their data, the methods used to collect it, and how it was processed to train the AI models. This will help prevent the introduction of biases into the AI system, which can lead to unfair or discriminatory outcomes.

In short, transparency and disclosure will play a vital role in ensuring the ethical and fair use of AI. This will require companies and organizations to be more open about the data and methods they use to train their AI systems, as well as the source code of their AI models. By doing so, we can build AI systems that are fair, unbiased, and truly serve the needs of society.

- **General Data Protection Regulation (GDPR)**: This is a regulation in the **European Union (EU)** that came into effect on May 25, 2018. GDPR sets out rules for how companies and organizations must protect the personal data of individuals in the EU. Today, GDPR regulations already enforce *"the right…to obtain an explanation of the decision reached"* by algorithms, and the EU has identified explainability as a key factor in increasing trust in AI in its white paper and AI regulation guidance. In my opinion, in addition to software disclosures, we also need control and a view of the hardware supply chain. We need a real-time view of both software and hardware supply chain systems to keep our trust and faith in digital computing systems.

- **AI-free zones**: Innovation across industries will bring software, digitization, and AI to business processes, from marketing to customer support. Some AI algorithms will make or affect decisions with direct and important consequences on people's lives. For instance, they diagnose medical conditions, screen candidates for jobs, process Visa applications, approve home loans, and decide on fines and jail sentences. In such circumstances, it may be wise to avoid using AI in some environments, which I call *AI-free zones*, or just let AI make a recommendation but leave the final decision to human judgment.

- **End user real-time feedback**: Today, most governments grapple with controlling misinformation or fake news spread on social media platforms. Social media companies have come up with ways where users can label and tag fake news that then gets triaged by social media companies. I think in the coming decades, governments will have stringent laws and recommendations that automatically force social media companies to take down any sensitive content. In the upcoming decades, AI will incorporate real-time feedback loops from users, government, and auto-triage systems to swiftly remove harmful content. Additionally, AI will automatically send messages to users who viewed the deleted content to provide updated information. This will help build trust, ethics, and awareness of what is and was shown to users on a social network.

As most AI and autonomous digital systems will work as independent entities, I think all systems must be created with mandatory options to stop them by the companies that own them, or be monitored and overseen by governments. Such steps will ensure both innovation and trust in technology. We should also have regulations that make technology companies that own and operate AI and automation technology both liable and accountable for actions taken by the software, bots, and devices. Such regulations will ensure researchers, developers, commercial operators of technology, and governments follow ethical practices. I like to call it this *ethical AI*, which ensures peace and harmony between machines, humans, and Mother Earth.

Summary

I hope and pray the world continues to bounce back from the COVID-19 pandemic. We have an amazing digital future ahead of us that was brought forward as a result of the pandemic. Every new invention or breakthrough will have its benefits and risks. As humans, we need to be aware of what we create and invent to make a better world and the associated risks. AI has massive advantages, and as we move forward in time, it will generate more jobs and inspire us to explore more around us, in space, across planets, and with better sustainable systems on mother earth. It's vital for us as humans to ensure that the rise of machines and AI in the coming decades doesn't get out of hand. I hope we learn from the COVID-19 pandemic and that our healthcare and research facilities are able to prevent future lockdowns. The investments and government focus that we have today give me comfort that, as a society, we are better equipped to fight the next pandemic. For sure, as a society, we are also becoming dependent on computerization and digitization so quickly that we are not equipped at all to protect ourselves from digital and AI-induced lockdowns. Technocrats, visionaries, and the developers of today have a real responsibility in ensuring that we build platforms and systems that have responsible and ethical AI systems to prevent a digital lockdown in the future.

Quite a lot rests on the shoulders of AI platform companies and AI developers to ensure that mankind is not faced with a permanent shutdown in the coming decades.

Further reading

- https://nickbostrom.com/ethics/artificial-intelligence.pdf
- https://www.vox.com/future-perfect/2018/12/21/18126576/ai-artificial-intelligence-machine-learning-safety-alignment
- https://builtin.com/artificial-intelligence/risks-of-artificial-intelligence

Index

packtpub.com

Subscribe to our online digital library for full access to over 7,000 books and videos, as well as industry leading tools to help you plan your personal development and advance your career. For more information, please visit our website.

Why subscribe?

- Spend less time learning and more time coding with practical eBooks and Videos from over 4,000 industry professionals

- Improve your learning with Skill Plans built especially for you

- Get a free eBook or video every month

- Fully searchable for easy access to vital information

- Copy and paste, print, and bookmark content

Did you know that Packt offers eBook versions of every book published, with PDF and ePub files available? You can upgrade to the eBook version at packtpub.com and as a print book customer, you are entitled to a discount on the eBook copy. Get in touch with us at customercare@packtpub.com for more details.

At www.packtpub.com, you can also read a collection of free technical articles, sign up for a range of free newsletters, and receive exclusive discounts and offers on Packt books and eBooks.

Other Books You May Enjoy

If you enjoyed this book, you may be interested in these other books by Packt:

Cybersecurity Leadership Demystified

Dr. Erdal Ozkaya

ISBN: 978-1-80181-928-2

- Understand the key requirements to become a successful CISO
- Explore the cybersecurity landscape and get to grips with end-to-end security operations
- Assimilate compliance standards, governance, and security frameworks
- Find out how to hire the right talent and manage hiring procedures and budget
- Document the approaches and processes for HR, compliance, and related domains
- Familiarize yourself with incident response, disaster recovery, and business continuity
- Get the hang of tasks and skills other than hardcore security operations

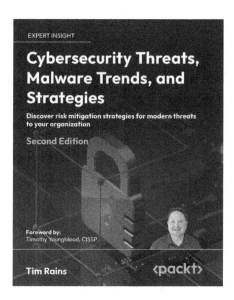

Cybersecurity Threats, Malware Trends, and Strategies - Second Edition

Tim Rains

ISBN: 978-1-80461-367-2

- Discover enterprise cybersecurity strategies and the ingredients critical to their success
- Improve vulnerability management by reducing risks and costs for your organization
- Mitigate internet-based threats such as drive-by download attacks and malware distribution sites
- Learn the roles that governments play in cybersecurity and how to mitigate government access to data
- Weigh the pros and cons of popular cybersecurity strategies such as Zero Trust, the Intrusion Kill Chain, and others
- Implement and then measure the outcome of a cybersecurity strategy
- Discover how the cloud can provide better security and compliance capabilities than on-premises IT environments

Packt is searching for authors like you

If you're interested in becoming an author for Packt, please visit `authors.packtpub.com` and apply today. We have worked with thousands of developers and tech professionals, just like you, to help them share their insight with the global tech community. You can make a general application, apply for a specific hot topic that we are recruiting an author for, or submit your own idea.

Share Your Thoughts

Now you've finished *Managing Risks in Digital Transformation*, we'd love to hear your thoughts! Scan the QR code below to go straight to the Amazon review page for this book and share your feedback or leave a review on the site that you purchased it from.

`https://packt.link/r/1803246510`

Your review is important to us and the tech community and will help us make sure we're delivering excellent quality content.

Download a free PDF copy of this book

Thanks for purchasing this book!

Do you like to read on the go but are unable to carry your print books everywhere? Is your eBook purchase not compatible with the device of your choice?

Don't worry, now with every Packt book you get a DRM-free PDF version of that book at no cost.

Read anywhere, any place, on any device. Search, copy, and paste code from your favorite technical books directly into your application.

The perks don't stop there, you can get exclusive access to discounts, newsletters, and great free content in your inbox daily

Follow these simple steps to get the benefits:

1. Scan the QR code or visit the link below

https://packt.link/free-ebook/9781803246512

2. Submit your proof of purchase

3. That's it! We'll send your free PDF and other benefits to your email directly

www.ingramcontent.com/pod-product-compliance
Lightning Source LLC
Chambersburg PA
CBHW082118070326
40690CB00049B/3612